Methods in Molecular Biology

Series Editor
John M. Walker
School of Life and Medical Sciences
University of Hertfordshire
Hatfield, Hertfordshire, UK

For further volumes:
http://www.springer.com/series/7651

For over 35 years, biological scientists have come to rely on the research protocols and methodologies in the critically acclaimed *Methods in Molecular Biology* series. The series was the first to introduce the step-by-step protocols approach that has become the standard in all biomedical protocol publishing. Each protocol is provided in readily-reproducible step-by-step fashion, opening with an introductory overview, a list of the materials and reagents needed to complete the experiment, and followed by a detailed procedure that is supported with a helpful notes section offering tips and tricks of the trade as well as troubleshooting advice. These hallmark features were introduced by series editor Dr. John Walker and constitute the key ingredient in each and every volume of the *Methods in Molecular Biology* series. Tested and trusted, comprehensive and reliable, all protocols from the series are indexed in PubMed.

Neurobiology

Methods and Protocols

Edited by

Sebastian Dworkin

Department of Microbiology, Anatomy, Physiology and Pharmacology, La Trobe University, Bundoora, VIC, Australia

Humana Press

Editor
Sebastian Dworkin
Department of Microbiology, Anatomy,
Physiology and Pharmacology
La Trobe University
Bundoora, VIC, Australia

ISSN 1064-3745 ISSN 1940-6029 (electronic)
Methods in Molecular Biology
ISBN 978-1-0716-3584-1 ISBN 978-1-0716-3585-8 (eBook)
https://doi.org/10.1007/978-1-0716-3585-8

This Humana imprint is published by the registered company Springer Science+Business Media, LLC, part of Springer Nature.
The registered company address is: 1 New York Plaza, New York, NY 10004, U.S.A.

Paper in this product is recyclable.

Preface

Most multi-cellular organisms comprise a defined "brain" region. The formation and development of this critical region, comprising closure of the neural tube, defined segmentation, neuronal differentiation, migration and integration, alongside a robust vascular network, requires an essential, integrated network of precisely regulated developmental cues and signals.

The study of neurobiology therefore necessitates a multitude of complementary approaches to fully understand how all the molecular and cellular components act together to correctly form the brain, and importantly, to also understand how the brain functions, preserves integrity following injury (stroke) and also how it ages.

The use of animal models as disparate as nematodes (*Caenorhabditis elegans*), zebrafish (*Danio rerio*) and mice (*Mus musculus*), as well as advanced imaging of the brain in human patients, has provided an incredible wealth of knowledge to allow us to understand how the human brain develops and functions, and importantly, also begin to understand what happens when things go wrong.

Ultimately, this book is intended to serve as a "go-to" guide for students in the field of neurobiology, whilst also acting as an essential guide to understanding neurobiology applications across disparate organisms and model systems. Moreover, we envision this book may also be of use to more senior researchers looking to expand their existing research programs to encompass novel techniques.

Bundoora, VIC, Australia ***Sebastian Dworkin***

Contents

Contributors

ERICKA ALLEN • *Flinders Health and Medical Research Institute, College of Medicine and Public Health, Flinders University, Bedford Park, SA, Australia*

SARAH JANE ANNESLEY • *Department of Microbiology, Anatomy, Physiology and Pharmacology, La Trobe University, Melbourne, VIC, Australia*

BENJAMIN HENRY ARNOLD • *Department of Microbiology, Anatomy, Physiology and Pharmacology, La Trobe University, Melbourne, VIC, Australia*

THIRUMA ARUMUGAM • *Department of Microbiology, Anatomy, Physiology and Pharmacology, Centre for Cardiovascular Biology and Disease Research, School of Agriculture, Biomedicine and Environment, La Trobe University, Bundoora, VIC, Australia*

M. BORKOWSKI • *Aurora Scientific, Aurora, ON, Canada*

YEE LIAN CHEW • *Flinders Health and Medical Research Institute, College of Medicine and Public Health, Flinders University, Bedford Park, SA, Australia*

J. E. CHURCH • *Aurora Scientific, Aurora, ON, Canada*

MICHELLE CORRONE • *School of Psychology and Public Health, La Trobe University, Melbourne, VIC, Australia*

SAMUEL HENRY CROSSMAN • *The Australian Regenerative Medicine Institute, Monash University, Clayton, VIC, Australia*

SALWADOR CYRANOWSKI • *Laboratory of Molecular Neurobiology, Nencki Institute of Experimental Biology, Polish Academy of Sciences, Warsaw, Poland; Postgraduate School of Molecular Medicine, Faculty of Medicine, Medical University of Warsaw, Warsaw, Poland*

T. MICHAEL DE SILVA • *Centre for Cardiovascular Biology and Disease Research, Department of Microbiology, Anatomy, Physiology and Pharmacology, La Trobe University, Bundoora, VIC, Australia*

MARIJA DINEVSKA • *Department of Surgery, Royal Melbourne Hospital, The University of Melbourne, Parkville, VIC, Australia*

QUYNH NHU DINH • *Department of Microbiology, Anatomy, Physiology and Pharmacology, Centre for Cardiovascular Biology and Disease Research, School of Agriculture, Biomedicine and Environment, La Trobe University, Bundoora, VIC, Australia*

SEBASTIAN DWORKIN • *Department of Microbiology, Anatomy, Physiology and Pharmacology, La Trobe University, Melbourne, VIC, Australia*

PAUL ROBERT FISHER • *Department of Microbiology, Anatomy, Physiology and Pharmacology, La Trobe University, Melbourne, VIC, Australia*

JARRAD FULLER • *Department of Microbiology, Anatomy, Physiology and Pharmacology, La Trobe University, Melbourne, VIC, Australia*

JEMMA GASPERONI • *Department of Microbiology, Anatomy, Physiology and Pharmacology, La Trobe University, Melbourne, VIC, Australia*

PATRICK GOODBOURN • *Melbourne School of Psychological Sciences, The University of Melbourne, Parkville, VIC, Australia*

EMILY J. JAEHNE • *School of Psychology and Public Health, La Trobe University, Melbourne, VIC, Australia*

TRAVIS K. JOHNSON • *School of Biological Sciences, Monash University, Clayton, VIC, Australia; Department of Biochemistry and Chemistry, and La Trobe Institute for Molecular Science, La Trobe University, Bundoora, VIC, Australia*

PATRICIA R. JUSUF • *School of Biosciences, The University of Melbourne, Parkville, VIC, Australia*

PAULINA KAMINSKA • *Laboratory of Molecular Neurobiology, Nencki Institute of Experimental Biology, Polish Academy of Sciences, Warsaw, Poland; Cellular Neurobiology Research Group, Faculty of Biology, University of Warsaw, Warsaw, Poland*

JAN KASLIN • *The Australian Regenerative Medicine Institute, Monash University, Clayton, VIC, Australia*

MITRA AMIRI KHABOOSHAN • *The Australian Regenerative Medicine Institute, Monash University, Clayton, VIC, Australia*

HYUN AH. KIM • *Centre for Cardiovascular Biology and Disease Research, Department of Microbiology, Anatomy, Physiology and Pharmacology, La Trobe University, Bundoora, VIC, Australia*

S. T. C. KUCEWICZ • *Department of Microbiology, Anatomy, Physiology & Pharmacology, School of Agriculture, Biomedicine & Environment, La Trobe University, Melbourne, VIC, Australia*

VERNISE J. T. LIM • *Department of Biochemistry and Chemistry, La Trobe Institute for Molecular Science, School of Agriculture, Biomedicine and Environment, La Trobe University, Bundoora, VIC, Australia*

JIAYI LIN • *School of Biological Sciences, Monash University, Clayton, VIC, Australia*

ANNA R. MALIK • *Cellular Neurobiology Research Group, Faculty of Biology, University of Warsaw, Warsaw, Poland*

THEO MANTAMADIOTIS • *Department of Surgery, Royal Melbourne Hospital, The University of Melbourne, Parkville, VIC, Australia; Department of Microbiology and Immunology, The University of Melbourne, Melbourne, VIC, Australia; Centre for Stem Cell Systems, The University of Melbourne, Parkville, VIC, Australia*

V. PRAGATHI MASAMSETTI • *Embryology Research Unit, Children's Medical Research Institute, University of Sydney, Westmead, NSW, Australia; School of Medical Sciences, Faculty of Medicine and Health, University of Sydney, Sydney, NSW, Australia*

DAIN L. MAXWELL • *Department of Biochemistry and Chemistry, La Trobe Institute for Molecular Science, School of Agriculture, Biomedicine and Environment, La Trobe University, Bundoora, VIC, Australia; Department of Anatomy and Neuroscience, The University of Melbourne, Parkville, VIC, Australia*

RILEY MCMAHON • *Embryology Research Unit, Children's Medical Research Institute, University of Sydney, Westmead, NSW, Australia; School of Medical Sciences, Faculty of Medicine and Health, University of Sydney, Sydney, NSW, Australia*

ANNA MCMILLEN • *Flinders Health and Medical Research Institute, College of Medicine and Public Health, Flinders University, Bedford Park, SA, Australia*

SARAH MELE • *School of Biological Sciences, Monash University, Clayton, VIC, Australia*

CAITLIN MINERVINI • *Flinders Health and Medical Research Institute, College of Medicine and Public Health, Flinders University, Bedford Park, SA, Australia*

JACQUELINE M. ORIAN • *Department of Biochemistry and Chemistry, La Trobe Institute for Molecular Science, School of Agriculture, Biomedicine and Environment, La Trobe University, Bundoora, VIC, Australia*

PAULINA PILANC • *Laboratory of Molecular Neurobiology, Nencki Institute of Experimental Biology, Polish Academy of Sciences, Warsaw, Poland*

MATTHEW D. W. PIPER • *School of Biological Sciences, Monash University, Clayton, VIC, Australia*

AELON RAHMANI • *Flinders Health and Medical Research Institute, College of Medicine and Public Health, Flinders University, Bedford Park, SA, Australia*

OANA SANISLAV • *Department of Microbiology, Anatomy, Physiology and Pharmacology, La Trobe University, Melbourne, VIC, Australia*

CHRISTOPHER G. SOBEY • *Centre for Cardiovascular Biology and Disease Research, Department of Microbiology, Anatomy, Physiology and Pharmacology, La Trobe University, Bundoora, VIC, Australia*

SEBASTIAN-ALEXANDER STAMATIS • *The Australian Regenerative Medicine Institute, Monash University, Clayton, VIC, Australia*

YICHEN SUN • *School of Life Science and Technology, The Key Laboratory of Developmental Genes and Human Disease, Southeast University, Nanjing, China; School of Biological Sciences, Monash University, Clayton, VIC, Australia*

TAMAR SZTAL • *School of Biological Sciences, Monash University, Melbourne, VIC, Australia*

PATRICK P. L. TAM • *Embryology Research Unit, Children's Medical Research Institute, University of Sydney, Westmead, NSW, Australia; School of Medical Sciences, Faculty of Medicine and Health, University of Sydney, Sydney, NSW, Australia*

MAARTEN VAN DEN BUUSE • *School of Psychology and Public Health, La Trobe University, Melbourne, VIC, Australia*

C. VAN DER POEL • *Department of Microbiology, Anatomy, Physiology & Pharmacology, School of Agriculture, Biomedicine & Environment, La Trobe University, Melbourne, VIC, Australia*

CELIA VANDESTADT • *The Australian Regenerative Medicine Institute, Monash University, Clayton, VIC, Australia*

SAMUEL S. WIDODO • *Department of Surgery, Royal Melbourne Hospital, The University of Melbourne, Parkville, VIC, Australia*

DAVID E. WONG ZHANG • *Centre for Cardiovascular Biology and Disease Research, Department of Microbiology, Anatomy, Physiology and Pharmacology, La Trobe University, Bundoora, VIC, Australia*

JIAHENG XIE • *School of Biosciences, The University of Melbourne, Parkville, VIC, Australia*

WEI XIE • *School of Life Science and Technology, The Key Laboratory of Developmental Genes and Human Disease, Southeast University, Nanjing, China*

SHENPENG R. ZHANG • *Centre for Cardiovascular Biology and Disease Research, Department of Microbiology, Anatomy, Physiology and Pharmacology, La Trobe University, Bundoora, VIC, Australia*

YU ZHAO • *School of Life Science and Technology, The Key Laboratory of Developmental Genes and Human Disease, Southeast University, Nanjing, China*

Chapter 1

Plate-Based Assays for the Characterization of Mitochondrial and Cellular Phenotypes

Benjamin Henry Arnold, Oana Sanislav, Paul Robert Fisher, and Sarah Jane Annesley

Abstract

The mitochondria are essential to eukaryotic life, acting as key drivers of energy generation while also being involved in the regulation of many cellular processes including apoptosis, cell proliferation, calcium homeostasis, and metabolism. Mitochondrial diseases which disrupt these processes lead to a diverse range of pathologies and lack consistency in symptom presentation. In disease, mitochondrial activity and energy homeostasis can be adapted to cellular requirements, and studies using *Dictyostelium* and human lymphoblastoid cell lines have shown that such changes can be facilitated by the key cellular and energy regulators, TORC1 and AMPK. Fluorescence-based assays are increasingly utilized to measure mitochondrial and cell signalling function in mitochondrial disease research. Here, we describe a streamlined method for the simultaneous measurement of mitochondrial mass, membrane potential, and reactive oxygen species production using MitoTracker Green™ FM, MitoTracker Red™ CMXRos, and DCFH-DA probes. This protocol has been adapted for both *Dictyostelium* and human lymphoblastoid cell lines. We also describe a method for assessing TORC1 and AMPK activity simultaneously in lymphoblastoid cells. These techniques allow for the characterization of mitochondrial defects in a rapid and easy to implement manner.

Key words Lymphoblast, *Dictyostelium*, ROS, Mitochondrial mass, TORC1, AMPK, Membrane potential

1 Introduction

The mitochondria are double-membrane-bounded organelles present in almost all human cell types. Believed to be of prokaryotic origin, the mitochondria have since been integrated into their eukaryotic hosts with most proteins involved in mitochondrial functions being encoded in the nuclear genome. However, the mitochondria retain a distinct, maternally inherited mtDNA genome, which in humans includes 13 protein-encoding genes as well as 22 tRNA genes [1]. While renowned for their roles as drivers of cellular energy generation by oxidative phosphorylation (OXPHOS), the mitochondria are also critical participants in other

Sebastian Dworkin (ed.), *Neurobiology: Methods and Protocols*, Methods in Molecular Biology, vol. 2746,
https://doi.org/10.1007/978-1-0716-3585-8_1,

cellular processes such as calcium homeostasis, cell death and proliferation, and other key biochemical processes including amino acid and lipid metabolism [2, 3].

ATP generation by mitochondrial OXPHOS accounts for approximately 90% of cellular ATP production, in a process where the flow of electrons from reducing substrates is coupled to the transfer of protons across the inner mitochondrial membrane by respiratory complexes I–IV, thereby maintaining an electrochemical gradient which can be consumed for ATP synthesis. Electrons are provided by redox substrates such as nicotinamide adenine dinucleotide (NAD^+), which is primarily generated at various stages in the tricarboxylic acid cycle in its reduced form (NADH) and is oxidized mainly by complex I to initiate respiration.

Electron transport across these respiratory complexes is also associated with the generation of most of the cell's reactive oxygen species (ROS) which are formed as electrons in the respiratory chain react with cellular oxygen. Elevated ROS levels may signal an increase in electron transfer through the ETC or downstream blockade of the electron transport process so that electron flow is diverted directly to O_2.

ROS are important modulators of numerous cellular processes and are involved in innate immune signalling and cell death induction [4, 5]. Despite this, ROS are damaging, and the effects of their generation in the vicinity of important OXPHOS machinery are typically constrained by ROS scavenging mechanisms. The capacity of these scavenging mechanisms may be exceeded by heightened ROS production associated with particular physiological conditions, and excess ROS generation has been linked to neuronal damage in neurodegenerative diseases [6].

Diseases associated with mitochondrial dysfunction are notably clinically diverse and can be categorized as either primary mitochondrial diseases, arising from impaired respiratory capabilities, or secondary mitochondrial diseases, which originate from non-respiratory mitochondrial defects [7]. Primary diseases can be further specified as those arising from defects in the nuclear mitochondrial genes, which are inherited in a Mendelian fashion, or those within the mtDNA, which are inherited maternally [8]. As hundreds of distinct mtDNA genomes can be present in the cell, mtDNA-encoded defects may only occur in a fraction of these genomes, leading to a state of "heteroplasmy" [9]. In this state, the presentation of clinical disease due to mitochondrial dysfunction may occur only once the accumulation of mutant genomes, or mutant load, exceeds a threshold level (usually 60–90%). Such a threshold may vary depending upon the mutation or the affected cell/tissue type in question. For example, neurogenic muscle weakness, ataxia, and retinitis pigmentosa (NARP) syndrome occur when ATPase 6 (m.8993T>G) mutational loads exceed 60%, while in tRNA mutants, the mutant load threshold for detecting a biochemical defect exceeds 90% [10, 11]. Cells with higher energy requirements may be more susceptible to mitochondrial dysfunction (e.g., neurons), whereas the effects of pathogenic mutations may be concealed in cells with lower energy demands [12]. As dynamic organelles, mitochondrial copy numbers are tightly regulated in order to fulfill physiological demands while limiting excess ROS generation. Cellular mitochondrial content (mitochondrial mass) is influenced by mitochondrial fission/fusion and biogenesis, a complex process that is influenced by central metabolic regulators, whose dysregulation has been associated with mitochondrial dysfunction in neurological diseases such as Alzheimer's and Parkinson's disease [13, 14].

Secondary mitochondrial disease is characterized by deficiency in the mitochondria's capacity for non-respiratory functions. Given the mitochondria's myriad roles, secondary mitochondrial diseases arising from mitochondrial dysfunction are unsurprisingly linked to a variety of physiologically diverse disorders [15]. For example, amyloid-β accumulation which occurs in Alzheimer's disease (AD) has been linked to alterations in mitochondrial dynamics, disturbances in which can attenuate the non-bioenergetic and respiratory activity of the mitochondria [16, 17]. Likewise, mitochondrial dysfunction in Parkinson's disease (PD) has been associated with altered mitochondrial dynamics, and the inhibition of excessive mitochondrial fission appeared to rescue the disease phenotype in a familial PD model [16, 18]. Inhibition of mitochondrial fission has also been shown to rescue cell viability and mitochondrial morphological defects in an amyotrophic lateral sclerosis (ALS) model, demonstrating the impact of secondary mitochondrial disease in perpetuating the abnormal physiological state in disease [19]. Previously overlooked, there is a growing understanding of the impact of secondary mitochondrial diseases, including those related to neuromuscular disorders, immunity, autism, cancer, and chromosomal defects [15]. This understanding will likely increase as the involvement of primary and secondary mitochondrial dysfunction in other diseases is investigated, highlighting the continuing need for reliable and informative assays of mitochondrial performance.

Regulatory control over cellular metabolism and mitochondrial dynamics is generally exerted through two essential protein complexes, the mechanistic target of rapamycin complex 1 (mTORC1 or simply TORC1) and AMP-activated protein kinase (AMPK). TORC1 is situated at the center of numerous signalling pathways which assess the availability of growth factors, oxygen, amino acids, and cellular energy stores [20, 21]. These signals in turn promote TORC1 activation, leading to a metabolic switch in favor of anabolic metabolism that promotes cell proliferation and growth [20, 21]. Additionally, mitochondrial gene expression is enhanced at the transcriptional and translational level by TORC1, upregulating mitochondrial biogenesis and increasing OXPHOS respiratory activity [13, 22].

Contrary to TORC1, the role of AMPK is that of a "cellular stress sensor" that can be activated by diverse cellular stresses but most notably "energy stress" which occurs when ATP levels decrease relative to those of AMP and ADP. AMPK activation typically serves to replenish cellular ATP by constraining excess energy expenditure and by increasing energy production through activation of glucose uptake and metabolism, mitochondrial biogenesis, and fatty acid metabolism [23, 24]. AMPK activation can restrict energy expenditure by inhibiting TORC1, either directly through TORC1's scaffolding subunit, Raptor, or indirectly through a heterodimeric upstream inhibitor of TORC1, namely, TSC1/TSC2 [25, 26]. Given their reciprocal roles as critical regulators of metabolism, chronic activation of either AMPK or TORC1 would lead to numerous detrimental consequences affecting crucial cellular functions. Thus, an understanding of the activation state of TORC1/AMPK would be informative when assessing the physiological consequences of mitochondrial disease. Combined with other "snapshot" measurements of mitochondrial activity, a detailed understanding of such consequences can begin to be developed.

Studies investigating mitochondrial function in disease, especially those with undetermined physiological mechanisms, are often limited by the physiological relevance of the cell type being studied. Consistent observations across multiple cell types however, such as those from peripheral tissues, can be revealing of pervasive systemic changes that would otherwise have remained concealed. For example, altered gene transcript and protein profiles appear consistent across ex vivo peripheral blood mononuclear cells (PBMCs) and those in the brain of neurodegenerative disease patients [27–29]. While studies using PBMCs can benefit from their noninvasive acquisition and ability to reflect systemic changes, they are metabolically quiescent, making it difficult to detect small changes in disease cohorts. An alternative is the use of immortalized lymphocytes known as lymphoblastoid cell lines (LCLs).

Generated by transfecting PBMCs (which are a mixture of circulating lymphocyte cell types) with Epstein-Barr virus (EBV), LCLs express a small number of EBV-associated proteins from a circular EBV episome which is required for their immortalization. These episomes are of low copy number (~20–30) and do not significantly alter the host genome [30, 31]. EBV transfection is selective for B cells, which then become stably proliferative and feature numerous advantages that make their use in mitochondrial studies more desirable than their lymphocyte progenitors. Unlike lymphocytes, LCL cultures can be maintained indefinitely, providing a virtually limitless supply of viable cells for study. LCLs also have higher oxidative requirements and contain higher mtDNA copy numbers than their lymphocyte counterparts, possibly due to their switch from a metabolically quiescent to an actively proliferating state [32]. As such, LCLs are well suited for use in research and are increasingly used to study mitochondrial function in a multitude of diseases [33] including Parkinson's disease [34], Huntington's disease [35], ALS [36], and myalgic encephalomyelitis/chronic fatigue syndrome [32]. For example, studies using PD-derived LCLs have shown that PD cells have elevated, not reduced, respiratory activity [34] which has since been confirmed in other cellular models [37–39]. Owing to their metabolically active state, LCLs may be more sensitive to changes in mitochondrial function than PBMCs and therefore more accurately reflect the underlying consequences of such changes, highlighting their value as models of mitochondrial disease.

In mammalian systems, the relationship between mutant genotypes and the presentation of mitochondrial disease is often ambiguous, owing to the interactions between gene- and individual-specific threshold effects, as well as varying levels of mitochondrial heteroplasmy. Model systems which lack these complications, such as the budding yeast *Saccharomyces cerevisiae*, the nematode *Caenorhabditis elegans*, and the *Drosophila* fly, are therefore well suited to studies concerning interactions between genetic mutations and mitochondrial dysfunction [40–42]. Another such model organism is the social eukaryotic amoeba, *Dictyostelium discoideum*. Recognized by the National Institutes of Health (www.nih.gov/science/models) as one of a small number of model organisms of recognized value for biomedical research, *D. discoideum* maintain many of the benefits found in other cellular models, such as having a sequenced, haploid genome that is amenable to manipulation. Uniquely, *D. discoideum* also conducts many of the cellular processes found in higher-order eukaryotes, such as cell motility and differentiation [43]. The *D. discoideum* life cycle comprises numerous stages and cellular arrangements which include unicellular propagation and aggregation, multicellular slug migration, and differentiation into fruiting bodies, thereby providing numerous measurable phenotypes which are readily accessible and reliable in presentation [44, 45]. These phenotypes are governed by multiple signalling pathways, some of which are involved in regulating metabolic functions, allowing for alterations in mitochondrial function to be associated with a reproducible pattern of phenotypic defects [46]. Mitochondrial disease modelling in *D. discoideum* has been accomplished using various methods, including depletion of mtDNA using ethidium bromide [47], gene knockdown of chaperonin 60 [48], and targeted gene disruption [46, 49–51]. These methods can create generalized mitochondrial defects as well as targeted defects in respiratory genes, thereby facilitating the modelling of primary mitochondrial disease.

Additionally, *D. discoideum* has been used to examine the nature of mitochondrial and biochemical dysfunction in secondary mitochondrial diseases [52] including the neurodegenerative conditions Alzheimer's disease [53–56], Parkinson's disease [38, 57–59], Huntington's disease [60], and Batten disease [61]. Clearly, *D. discoideum* exhibit numerous qualities that make its use in mitochondrial disease modelling advantageous, as evidenced by its demonstrated history as a model for studying both primary and secondary mitochondrial disorders [52].

Here, we present a method for rapidly assessing various parameters of mitochondrial function in *D. discoideum* and lymphoblastoid cell lines using simple plate-based assays. These parameters include measurements of mitochondrial mass, mitochondrial membrane potential (MMP), and ROS generation. Furthermore, we describe methods for assaying total cellular NAD levels as well as measuring TORC1 and AMPK activation in LCLs.

Typical plate-based assays provide a "snapshot," steady-state measurement of particular mitochondrial parameters. These readouts are ideally used in conjunction with other measurements of mitochondrial function, such as Seahorse respirometric flux assays which provide real-time metabolic rate measurements and have previously been adapted for use in *D. discoideum* and LCLs [32, 62].

2 Materials

2.1 Cell Culture

Dictyostelium cells grown in HL-5 medium (*see* **Note 1**) at 21 °C in shaken suspension, to exponential phase with a cell density ranging from 0.8 to 3×10^6 cells/mL. Parental wildtype cell lines should be used as a control cell line when assessing mutant cell lines. The cells are grown without antibiotic selection for several generations before starting the assays, to ensure any differences observed are due to the mutations and not caused by the presence of antibiotics.

1. Human lymphoblastoid cell lines are grown in Minimal Essential Medium α (MEMα) supplemented with 10% fetal bovine serum (FBS) and 1% penicillin-streptomycin (Pen-Strep) in a 37 °C humidified incubator with 5% CO_2. Cells are subcultured generally every 2–3 days, by replacing one third up to a half of the volume of cells/media with fresh, pre-warmed MEMα. Cells should be cultured for approximately 1 week after thawing before being used in assays.

2.2 Equipment

1. 96-well plates: Clear.
2. 96-well plates: Black-walled, clear bottom. For fluorescence measurements of mitochondrial mass, membrane potential, and reactive oxygen species (ROS). Either white or black plates can be used here.

3. 384-well microplates: Low volume, white-walled, white bottom for fluorescence resonance energy transfer (FRET)-based AMPK and TORC1 assays and NAD/NADH assays.
4. 384-well plates: Clear, flat bottomed, sterile, tissue culture grade. For treating cell lines with pharmaceutical agents and reagents.
5. Transparent plastic sealing film for polymerase chain reaction (PCR) 96-well plates.
6. CLARIOstar plate reader or any plate reader that can conduct fluorescent, luminescent, and homogeneous time-resolved fluorescence (HTRF) measurements.
7. Horizontal orbital shakers.
8. 1.5 mL microcentrifuge tubes.
9. CO_2 incubator.
10. Hemocytometer or automated cell counter.

2.3 Reagents and Kits

The following reagents, where appropriate, are best stored in small volume aliquots to prevent excessive freeze-thaw cycles as well as to prevent their unnecessary exposure to light, which may affect their fluorescence during testing:

1. MitoTracker Red™ CMXRos dye: Create a 200 μM stock solution by diluting the vial contents (50 μg) with 470.35 μL dimethyl sulfoxide (DMSO). Store 20 μL small volume aliquots at −20 °C. Excitation 579 nm, emission 599 nm (measured with the TEXAS red filter).
2. MitoTracker Green™ FM dye: Create a 400 μM stock solution by diluting the vial contents (50 μg) with 186.05 μL DMSO. Store 20 μL small volume aliquots at −20 °C. Excitation 490 nm, emission 516 nm (measured with the FITC filter).
3. Dichlorodihydrofluorescein diacetate (DCFH-DA) for ROS detection: Prepare a 25 mM solution with distilled H_2O, and store in small volume 20 μL aliquots at −20 °C. Excitation 504 nm, emission 524 nm (measured with the FITC filter).
4. NAD/NADH Glo assay kit, comprising: Reductase, reductase substrate, NAD cycling enzyme (lyophilized), NAD cycling substrate, luciferin detection reagent, and reconstitution buffer.
5. Phospho-4E-BP1 (Thr37/46) cellular assay kit: Store at −80 °C.
6. Phospho-ACC (Ser79) cellular assay kit: Store at −80 °C.
7. Torin2, TORC1 inhibitor: Prepare a 500 μM stock solution by adding 4.62 mL of DMSO to 1 mg of Torin2 powder. Store 200 μL aliquots at −20 °C for long term or 4 °C for short-term storage.

8. Compound C (Dorsomorphin 2HCl), AMPK inhibitor: Prepare a 10 mM stock solution by adding 250 μL DMSO to 1 mg powder. Store in 50 μL aliquots at −20 °C.
9. A769662, AMPK activator: Prepare a 50 mM stock by diluting the vial contents (10 mg) with 555 μL DMSO. Store in small volume aliquots (20 μL) at −20 °C.
10. Trypan blue, 0.4% solution.
11. Hoechst 33342, trihydrochloride: Prepare a 2 mg/mL stock solution by diluting the vial contents with DMSO. Store at − 20 °C. Excitation 361 nm, emission 497 nm, fluorescent signal usually measured with the 4′,6-diamidino-2-phenylindole (DAPI) filter.
12. HL-5 medium.
13. Minimal Essential Medium α (MEMα) supplemented with 10% FBS and 1% penicillin-streptomycin (Pen-Strep).
14. Low fluorescence (Lo-Flo) HL-5 medium for the *Dictyostelium* experiments: Add 1.6 g proteose peptone, 0.45 g yeast extract, 0.484 g KH_2PO_4, and 1.2 g $Na_2HPO_4.12H_2O$ per L of distilled water. Adjust the pH to 6.5 with NaOH and filter sterilize. The media can be stored at room temperature or at 4 °C but should be warmed to 22 °C before use in experiments.
15. Phosphate-buffered saline (PBS): Add 8 g NaCl to 800 mL distilled H_2O, and then add 0.187 g KCl and 1.42 g Na_2HPO_4. Adjust the pH to 7.4 with HCl, and fill to a total volume of 1 L with distilled H_2O.

2.4 Mitochondrial Staining Solutions

The volumes and amounts referred to below are sufficient for the testing of up to 12 different cell lines/strains in a full 96-well plate. Volumes can be adjusted to the number of wells or cell lines used. Once prepared, staining solutions should be protected from light sources, kept at 4 °C, and prepared fresh for each experiment:

1. MitoTracker Red™ CMXRos staining solution: Prepare 2600 μL of a 400 nM stock solution by adding 5.2 μL MitoTracker Red™ CMXRos 200 μM stock solution into 2600 μL of Lo-Flo HL-5 medium and shake gently.
2. MitoTracker Green™ FM staining solution: Prepare 2600 μL of an 800 nM stock solution by adding 5.2 μL MitoTracker Green™ 400 μM stock solution into 2600 μL of Lo-Flo HL-5 medium and shake gently.
3. DCFH-DA staining solution: Prepare 2600 μL of a 25 μM stock solution by diluting 2.6 μL DCFH-DA 25 mM stock solution with 2600 μL of Lo-Flo HL-5 medium.

4. To measure background fluorescence, prepare a control solution by dissolving the same amount of DMSO (5.2 μL) in 2600 μL of Lo-Flo HL-5 medium.

2.5 NAD/NADH Detection Reagents

Preparation of reagents:

1. Reconstituted luciferin detection reagent: Equilibrate the reconstitution buffer and the lyophilized luciferin detection reagent to RT, and then transfer the entire volume of the reconstitution buffer into the lyophilized luciferin detection reagent and mix by inverting 4–6 times. Aliquot into single-use volumes (1 mL) and store at −20 °C.
2. Equilibrate all reagents on ice.
3. To prepare 1040 μL of NAD-Glo detection reagent, add 1 mL of reconstituted luciferin detection reagent, 5 μL of reductase, 5 μL of reductase substrate, 5 μL of NAD cycling enzyme, and 25 μL of NAD cycling substrate. Mix gently by inverting 4–6 times and keep all items at room temperature, covered, until use. These volumes should be adjusted depending on the number of plates or volumes used (*see* **Note 2**).

2.6 AMPK and TORC1 Activity Assay Reagents

Assay reagents should be prepared prior to the assay (*see* **Note 3**). Preparation of reagents:

1. 4E-BP1 assay of TORC1 activity:
   - Prepare a DMSO vehicle control by adding 3 μL of DMSO to 497 μL of MEMα (*see* **Note 4**).
   - Prepare a 3 μM Torin2 working stock by diluting 500 μM Torin2 stock with 497 μL MEMα.
2. ACC assay of AMPK activity:
   - Prepare a DMSO vehicle control by diluting 3 μL of DMSO with 497 μL of MEMα.
   - Prepare a 60 μM Compound C working stock by diluting 3 μL of 10 mM Compound C stock with 497 μL of MEMα.
   - Prepare a 180 μM A769662 working stock by diluting 1.8 μL of 50 mM A769662 stock with 498.2 μL MEMα.

3 Methods

The reagents DCFH-DA, MitoTracker Red™ CMXRos, and MitoTracker Green™ FM (*see* **Notes 5–7**) used to measure ROS, MMP, and mitochondrial mass are applicable to both human LCLs and *D. discoideum* cells (*see* **Note 8**):

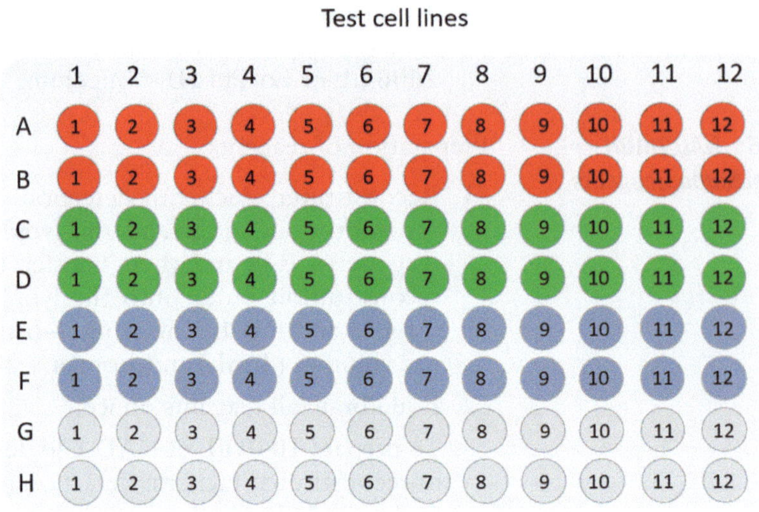

Fig. 1 Example plate layout for measuring mitochondrial mass, MMP, and ROS using a single plate. Each column is occupied by cells from a single cell line and indicated by the numbers in the wells. Mitochondrial dyes are added to each row, in duplicate. In this example, MitoTracker Red™ CMXRos is added to rows A and B (red wells), MitoTracker Green™ FM is added to rows C and D (green wells), DCFH-DA is added to rows E and F (blue wells), and DMSO/LoFlo HL-5 background control medium is added to rows G and H (gray wells). Fluorescence is measured as follows: green with excitation 483 nm and emission 530 nm, red with excitation 570 nm and emission 630, and blue with excitation 360 nm and emission 490 nm. Measurements are then used in subsequent calculations as described in the text

3.1 Mitochondrial Mass, Mitochondrial Membrane Potential, and Reactive Oxygen Species Measurements in Dictyostelium *Cells*

1. Count *D. discoideum* cells grown to early exponential phase (1–2.5 × 10^6 cells/mL), using a hemocytometer. Harvest a total of 1 × 10^6 cells by centrifugation at 1000×*g* for 2 min, and carefully remove the supernatant. Resuspend the cell pellet in 1 mL room temperature Lo-Flo-HL-5 medium. Seed 100 μL (10^5 cells) in each well of a 96-well, white, flat-bottom plate, using a total of 8 wells in 1 column for each cell line/strain (Fig. 1) (*see* **Note 9**).
2. Incubate the plate at room temperature for 30 min.
3. Carefully add 100 μL of the prepared mitochondrial staining solutions: MitoTracker Red™ CMXRos to rows A and B and MitoTracker Green™ FM to rows C and D (Fig. 1). Add 100 μL of DCFH-DA staining solution to rows E and F. Add 100 μL of Lo-Flo HL-5 medium containing 0.2% DMSO (*see* **Note 10**) to the remaining rows (Rows G and H) to control for background fluorescence.
4. Incubate the plate in the dark, at room temperature for 40 min (*see* **Note 11**). Measure the fluorescent signal on a microplate reader using the following setup:

- Top optic
- Settling time: 0.2 s.
- Focal height dependent on the volumes and types of plates used (*see* **Note 12**).
- Well reading method: Well scan with spiral averaging.
- Number of flashes per well: 70.
- Measure 2 fluorophores: Green with excitation 483 nm, emission 530 nm, gain 700. Red with excitation 570 nm, emission 630, gain 1500 (*see* **Note 13**).

5. Save results for data analysis.
6. Calculations:
   (a) Calculate mitochondrial mass by subtracting the background green fluorescence signal (cells with no dye, rows G and H) from the sample green fluorescence signal (cells with dye, rows C and D).
   (b) Calculate MitoTracker Red™ CMXRos signal by subtracting the background red fluorescence signal (rows G and H) from the sample red fluorescence signal (rows A and B).
   (c) MMP is the ratio of the background-subtracted MitoTracker Red™ CMXRos signal to the background-subtracted MitoTracker Green™ FM signal.
   (d) Calculate ROS levels by subtracting the background green fluorescence (rows G and H) from the sample DCFH green fluorescence (rows E and F).

3.2 Mitochondrial Mass, Mitochondrial Membrane Potential, and Reactive Oxygen Species Measurements in Lymphoblastoid Cell Lines

1. Harvest approximately 5 mL of each confluent culture to be tested by centrifugation at 500 ×*g* for 5 min. Pour off the media, and resuspend the cell pellet in 1 mL fresh MEMα.
2. Measure cell density by trypan blue staining: Mix 10 μL of trypan blue with 10 μL of cell suspension, and count using a hemocytometer (*see* **Note 14**).
3. Adjust each cell suspension to 1×10^6 cells/mL by adding fresh MEMα. Seed 100 μL (1×10^5 cells) in each well of a 96-well, white, flat-bottom plate, using a total of 8 wells in 1 column for each cell line.
4. Incubate the plate at 37 °C for 30 min.
5. Add 2.6 μL Hoechst 33342 nuclear stain in DMSO into 5.2 mL of the pre-prepared staining solutions (MitoTracker Red™ CMXRos, MitoTracker Green™ FM, and DCFH-DA described in Subheading 2.4) (*see* **Note 15**). Carefully add 100 μL of each prepared staining solution to each well: MitoTracker Red™ CMXRos to rows A and B, Mitotracker

Green™ FM to rows C and D, and DCFH-DA staining solution to rows E and F (Fig. 1). To the remaining wells (Rows G and H), add 100 μL of MEMα containing 0.3% DMSO (without Hoechst nuclear stain) to act as a background control (*see* **Note 10**).

6. Incubate the plate at 37 °C in the dark for 40 min.
7. Measure the fluorescent signal on a microplate reader using the following setup:
   - Top optic.
   - Settling time: 0.2 s.
   - Focal height dependent on the volumes and types of plates used (*see* **Note 11**).
   - Well reading method: Well scan with spiral averaging.
   - Number of flashes per well: 70.
   - Measure 3 fluorescences: Green with excitation 483 nm, emission 530 nm, gain 700. Red with excitation 570 nm, emission 630, gain 1500. Blue with excitation 360 nm, emission 490 nm, gain 1100 (*see* **Note 13**).
8. Calculations:
   (a) Calculate mitochondrial mass (normalized MitoTracker Green™ FM signal) by first calculating the ratio of the background-subtracted green fluorescence (average of (Rows C and D)—average of (Rows G and H)) to the nuclear staining signal (as a proxy measure of total cell mass), which is the background-subtracted blue fluorescence (average of (Rows C and D)—average of (Rows G and H)).

   (b) Calculate the normalized MitoTracker Red™ CMXRos signal by first calculating the ratio of the background-subtracted red fluorescence signal (average of (Rows A and B)—average of (Rows G and H)) to the nuclear staining signal measured as the background-subtracted blue fluorescence (average of (Rows C and D)—average of (Rows G and H)).

   (c) Calculate MMP by calculating the ratio of the normalized MitoTracker Red™ CMXRos signal (Step 6b) to the normalized MitoTracker Green™ FM signal (Step 6a).

   (d) Calculate ROS levels by calculating the ratio of the background-subtracted DCFH green signal (average of (rows E and F)—average of (Rows G and H)) to the background-subtracted blue fluorescent nuclear staining signal (average of (Rows E and F)—average of (Rows G and H)).

3.3 Total Cellular NAD Assay

A luciferase-based assay is described which couples NADH regeneration with the reduction of proluciferin to luciferin, whose fluorescence can then be measured. The following protocol can be used for both *Dictyostelium* and human lymphoblastoid cell lines:

1. Harvest 3×10^5 cells for each cell line by centrifugation (500×*g* for 5 min for LCL, 1000×*g* for 2 min for *Dictyostelium* cells). Resuspend the cell pellet in 100 μL PBS.
2. Seed 10 μL of cells (3×10^4 cells) per well in a 384-well, white, low volume plate. Include duplicate wells for each cell line.
3. Add 10 μL NAD-Glo detection reagent (Subheading 2.4) to each of the assay wells.
4. Using an orbital shaker, shake the plate for 3–5 s (100 rpm) to mix the cell suspension and NAD-Glo detection reagent. Take care to ensure that this is gentle (*see* **Note 16**).
5. Seal the plate with a clear plate seal cover, and incubate in the dark for 1 h.
6. Measure luminescence on a microplate reader.

3.4 AMPK and TORC1 Activity Assays for Human LCLs

Here, we describe a method for measuring TORC1 and AMPK activity by measuring the phosphorylation state of their respective downstream targets, namely, 4E-BP1 and ACC. The Phospho-4E-BP1 (Thr37/46) cellular assay kit and Phospho-ACC (Ser79) cellular assay kits are homogeneous time-resolved fluorescence (HTRF)-based assays, which involve a combination of phospho-specific and protein-specific antibodies. In such assays, one antibody features a donor fluorophore (cryptate) and the other an acceptor (D2), such that the proximity between phospho-specific and protein-specific antibodies after binding generates a fluorescence resonance energy transfer (FRET) signal. The strength of this signal is proportional to the amount of phosphorylated protein, which is calculated by comparing D2 acceptor emission at 665 nm to the emission of the cryptate donor at 620 nm. This comparison is made as the cryptate donor antibody will fluoresce regardless of whether a FRET signal has been generated, while the acceptor will only fluoresce at 665 nm once a FRET signal has been generated:

1. Harvest approximately 5 mL of confluent cell culture by centrifugation at 500×*g* for 5 min. Pour off the media and resuspend the cell pellet in 500 μL fresh MEMα.
2. Assess cell density with trypan blue staining: Mix 10 μL of trypan blue with 10 μL of cell suspension, and count using a hemocytometer.
3. Transfer 5.5×10^5 cells to a clean microcentrifuge tube. Fill the tube to a total volume of 275 μL using fresh MEMα.

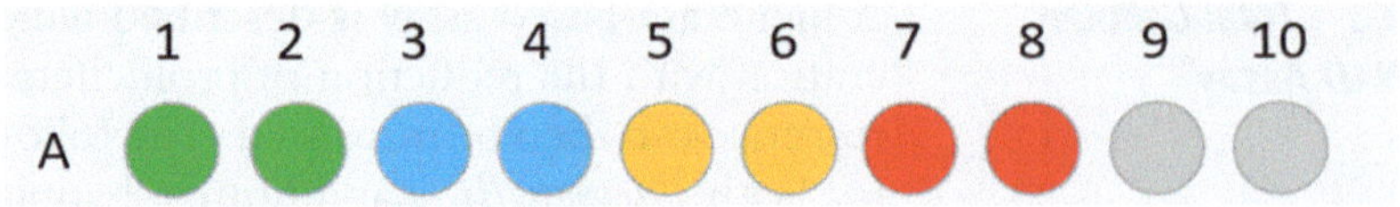

Fig. 2 Example plate layout for assaying AMPK and TORC1 activity. Cell lysates from single cell lines/strains are added to each row. Four wells are required for assaying TORC1 activity, and six wells are required for assaying AMPK activity. The green wells are allocated for the 4E-BP1 assay DMSO control, and the cyan wells are allocated for TORIN2 (an inhibitor of TORC1) treatment. The orange wells are allocated for the ACC assay DMSO control, the red cells are allocated for Compound C treatment (AMPK inhibitor), and the gray wells are allocated for A769662 treatment (AMPK activator). Duplicate wells are used for each group

4. Plate 25 μL (5×10^4 cells) of cell suspension per well in a clear 384-well plate. Allocate 10 wells for each cell line:
   - 4 wells for the phospho-4E-BP1 assay (2 wells with Torin2 treatment, 2 wells for untreated DMSO vehicle controls) (Fig. 2).
   - 6 wells for the phospho-ACC assay (2 wells with AMPK inhibitor Compound C treatment, 2 wells with AMPK activator A769662 treatment, and 2 wells for untreated DMSO vehicle controls) (Fig. 2) (*see* **Note 17**).
5. Aliquot 5 μL of the previously prepared Torin2, Compound C, and A769662 working stock solutions (Subheading 2.6) into the wells of a 384-well plate (10 wells per cell line) (Fig. 2).
6. Incubate the plate at 37 °C with 5% CO_2 for 3 h.
7. While the plate is incubating, prepare separate lysis buffers for each assay. Calculate the required amounts of each buffer necessary for the assay. For a single cell line, 40 μL of 4E-BP1 lysis buffer and 60 μL of ACC lysis buffer will be required (*see* **Note 18**). Defrost the blocking buffers and dilute 1:25 in lysis buffer (e.g., 10 μL of blocking buffer will need 240 μL of lysis buffer) to create supplemented lysis buffer.
8. After incubation, add 10 μL of supplemented lysis buffer to each well. Ensure that the supplemented lysis buffers for 4E-BP1 and ACC are added to the correct wells (Fig. 2).
9. Cover the plate, and then incubate at room temperature for 40 min with shaking (*see* **Note 19**).
10. Prepare antibodies in detection reagent: Defrost antibodies on ice (4E-BP1 D2 and 4E-BP1 cryptate antibodies for the TORC1 assay, ACC D2 and ACC cryptate antibodies for the AMPK assay). Prepare each of the four antibody mixtures in separate labelled microcentrifuge tubes, by diluting each antibody 1:20 with detection reagent. For each cell line, 8 μL of

the 4E-BP1 antibodies (2 μL/well) and 12 μL of the ACC antibodies (2 μL/well) is required.

11. Prepare a detection plate using a new 384-well, white, small volume plate:
    (a) Prepare a blank control well for each assay, containing 16 μL of cell lysate, 2 μL of detection buffer, and 2 μL of diluted D2 antibody mixture only. This is used to check for unspecific cryptate signal.
    (b) Prepare one negative control containing 16 μL MEMα for each assay with no cells to check for unspecific signal.
    (c) Prepare one positive control well for each assay, containing 16 μL of control lysate included in the assay kit.
    (d) Mix the D2 and cryptate antibody solutions prepared in Step 10 together. There should now be a 4E-BP1 antibody mixture and an ACC antibody mixture.
    (e) Add 4 μL of the antibody mixture prepared above to each experimental well, and the prepared negative and positive controls. Do not add any antibody mixture to the blank control well.
12. Cover the plate with a plate sealer and incubate at room temperature for 24 h in darkness.
13. Scan the plate on a CLARIOstar or other HTRF-compatible plate reader for Eu3+ cryptate with emission at 2 wavelengths: 665 nm and 620 nm. Make sure the plastic cover is removed before scanning. Settings are as follows:
    - Top optic.
    - Integration time: 60 μs.
    - Time: 400 μs.
    - Settling time: 0.1 s.
    - Number of flashes per well: 200.
    - Gain for 665 nm: 2500.
    - Gain for 620 nm: 2300.
14. Calculate the ratio between the 2 signals for each plate:

$$\text{FRET ratio} = \frac{\text{Signal 665 nm}}{\text{Signal 620 nm}} \times 10^4$$

4 Notes

1. HL-5 medium can also be prepared by mixing 10 g proteose peptone, 5 g yeast extract, 0.43 g $Na_2HPO_4 \times 2H_2O$, 0.35 g

KH_2PO4, and 10 g glucose. The pH of the media is adjusted to 6.4–6.6 with glacial acetic acid and sterilized by autoclaving.

2. The NAD/NADH Glo detection reagent should be prepared immediately prior to assay setup and used within a few hours. Do not store any remaining detection reagent.
3. 4E-BP1 and ACC phosphorylation assays for measuring TORC1 and AMPK activity are usually performed simultaneously on a 384-well plate.
4. We use DMSO-treated cells here, but untreated cells could also be included in the assay.
5. Mitochondrial mass is measured using a fluorescent dye, MitoTracker Green™ FM, which is a mitochondrial staining solution that accumulates in the mitochondrial inner membrane, binding to a subset of mitochondrial proteins with free thiol groups contained in cysteine residues [63]. Importantly, the binding behavior of MitoTracker Green™ FM is unaffected by MMP.
6. To measure MMP, two mitochondrial staining solutions are used, the first being MitoTracker RedTM CMXRos. MitoTracker RedTM CMXRos features a chloromethyl group which can form thioester bonds with inner membrane mitochondrial thiol groups, and its fluorescence is positively affected by elevations in MMP [64]. Because the MitoTracker Red signal also depends upon the total amount of active mitochondrial membrane, its fluorescence must be expressed relative to the MitoTracker GreenTM FM signal, which is unaffected by MMP [65].
7. DCFH-DA can freely diffuse into cells, where it is deacetylated by cellular esterases into a nonfluorescent compound that can be oxidized by ROS into the fluorescent 2′,7′--dichlorofluorescein (DCF) [66]. Thus, ROS generation in DCFH-DA treated cells creates a measurable fluorescent signal which correlates with ROS production.
8. We typically perform these three types of assays simultaneously, making it easier to compare changes and differences between cell lines.
9. Despite the fact that black plates are recommended for fluorescent assays, we have noticed that white plates have provided a better resolution when measuring MitoTracker Green™ FM signal.
10. This is the maximum concentration of DMSO in our prepared mitochondrial staining solutions, so the background wells are dosed to reflect this amount. Should the dyes be prepared differently, the background wells should be adjusted to reflect these updated conditions.

11. The incubation time can vary between 30 and 60 min, depending on the strain.
12. Focal height settings on the microplate reader will need to be adjusted as necessary: in most cases to between 6.7 and 7.6 mm.
13. Gain can be adjusted as required to prevent signal overflow.
14. Alternatively, an automated cell counter, such as a Countess 2 cell counter, can be used here.
15. An additional reagent, Hoechst 33342 nuclear stain, is used in the human LCL protocol as a necessary normalization measurement. This step is not included in the *D. discoideum* protocol as they are resistant to staining by Hoechst 3342 nuclear stain.
16. We have noticed that the experiment may fail when the plate is shaken too vigorously, possibly due to cellular damage releasing contents which could inactivate enzymes in the assay.
17. Here, we use compounds that are inhibitory to TORC1 (Torin2) and AMPK (Compound C) as well as an AMPK activator (A769662) to assess the activity of these complexes more comprehensively. This allows changes in the drug-inhibitable/activatable fractions of the phosphorylation signal to be measured, which may be more specific or better reflect other changes in the kinase/phosphatase signalling networks involved.
18. 40 μL of 4E-BP1 lysis buffer and 60 μL of ACC lysis buffer are the *minimum* volumes required for a single cell line. To account for pipetting error, the volume of lysis buffer should be slightly increased. Take care to account for this increase in volume when supplementing the lysis buffers with blocking buffer.
19. At this point in the assay, the plate can be stored at −80 °C.

References

1. Anderson S, Bankier AT, Barrell BG, de Bruijn MHL, Coulson AR, Drouin J, Eperon IC, Nierlich DP, Roe BA, Sanger F, Schreier PH, Smith AJH, Staden R, Young IG (1981) Sequence and organization of the human mitochondrial genome. Nature 290:457–465
2. Bertero OE, Maack C (2018) Calcium signalling and reactive oxygen species in mitochondria. Circ Res 122(10):1460–1478
3. Vakifahmetoglu-Norberg H, Ouchida AT, Norberg E (2017) The role of mitochondria in metabolism and cell death. Biochem Biophys Res Commun 482:426–431
4. West AP, Shadel GS, Ghosh S (2011) Mitochondria in innate immune responses. Nat Rev Immunol 11(6):389–402
5. Marchi S, Giorgi C, Suski JM, Agnoletto C, Bononi A, Bonora M, De Marchi E, Missiroli S, Patergnani S, Poletti F, Rimessi A, Duszynski J, Wieckowski MR, Pinton P (2012) Mitochondria-ROS crosstalk in the control of cell death and aging. J Signal Transduct 2012: 329635
6. Kausar S, Wang F, Cui H (2018) The role of mitochondrial in reactive oxygen species generation and its implications for neurodegenerative diseases. Cell 7:274

7. Alston CL, Rocha MC, Lax NZ, Turnbull DM, Taylor RW (2017) The genetics and pathology of mitochondrial disease. J Pathol 241:236–250
8. Giles RE, Blanc H, Cann HM, Wallace DC (1980) Maternal inheritance of human mitochondrial DNA. Proc Natl Acad Sci U S A 77(11):6715–6719
9. Stefano GB, Kream RM (2016) Mitochondrial DNA heteroplasmy in human health and disease. Biomed Rep 4(3):259–262
10. Mäkelä-Bengs P, Suomalainen A, Majander A, Rapola J, Kalimo H, Nuutila A, Pihko H (1995) Correlation between the clinical symptoms and the proportion of mitochondrial DNA carrying the 8993 point mutation in the NARP syndrome. Pediatr Res 37(5):634–639
11. Yoneda M, Miyatake T, Attardi G (1995) Heteroplasmic mitochondrial tRNA(lys) mutation and its complementation in MERRF patient-derived mitochondrial transformants. Muscle Nerve 18(S14):95–101
12. Zhao XY, Lu MH, Yan DJ, Xu DE, Yao PP, Ji WL, Chen H, Liu WL, Yan CX, Xia YY, Li S, Tao J, Ma QH (2019) Mitochondrial dysfunction in neural injury. Front Neurosci 13:30
13. Cunningham JT, Rodgers JT, Arlow DH, Vazquez F, Mootha VK, Puigserver P (2007) mTOR control mitochondrial oxidative function through a YY1-PGC-1a transcriptional complex. Nature 450:736–741
14. Sheng B, Wang X, Su B, Lee HG, Casadesus G, Perry G, Zhu X (2012) Impaired mitochondrial biogenesis contributes to mitochondrial dysfunction in Alzheimer's disease. J Neurochem 120:419–429
15. Niyazov DM, Kahler SG, Frye RE (2016) Primary mitochondrial disease and secondary mitochondrial dysfunction: importance of distinction for diagnosis and treatment. Mol Syndromol 7:122–137
16. Compagnoni GM, Di Fonzo A, Corti S, Comi GP, Bresolin N, Masliah E (2020) The role of mitochondria in neurodegenerative diseases: the lesson from Alzheimer's disease and Parkinson's disease. Mol Neurobiol 57(7): 2959–2980
17. Archer SL (2013) Mitochondrial dynamics – mitochondrial fission and fusion in human diseases. N Engl J Med 369:2236–2251
18. Cui M, Tang X, Christian WV, Yoon Y, Tieu K (2010) Perturbations in mitochondrial dynamics induced by human mutant PINK1 can be rescued by the mitochondrial division inhibitor mdivi-1. J Biol Chem 285(15):11740–11752
19. Song W, Song Y, Kincaid B, Bossy B, Bossy-Wetzel E (2013) Mutant SOD1^{G93A} triggers mitochondrial fragmentation in spinal cord motor neurons: neuroprotection by SIRT3 and PGC-1α. Neurobiol Dis 51:72–81
20. Laplante M, Sabatini DM (2012) mTOR signaling in growth control and disease. Cell 149: 274–293
21. Howell JJ, Ricoult SJH, Ben-Sahara I, Manning BD (2013) A growing role for mTOR in promoting anabolic metabolism. Biochem Soc Trans 41:906–912
22. Morita M, Gravel SP, Chènard V, Sikström K, Zheng L, Alain T, Gandin V, Avizonis D, Arguello M, Zakaria C, McLaughlan S, Nouet Y, Pause A, Pollak M, Gottlieb E, Larsson O, St-Pierre J, Topisirovic I, Sonenberg N (2013) mTORC1 controls mitochondrial activity and biogenesis through 4E-BP-dependent translational regulation. Cell Metab 18:698–711
23. Marsin AS, Bouzin C, Bertrand L, Hue L (2002) Stimulation of glycolysis by hypoxia in activated monocytes is mediated by AMP-activated protein kinase and inducible 6-phosphofructo-2-kinase. J Biol Chem 277: 30778–30783
24. Jäger S, Handschin C, St. Pierre J, Spiegelman BM (2007) AMP-activated protein kinase (AMPK) action in skeletal muscle via direct phosphorylation of PGC-1a. Proc Natl Acad Sci U S A 104:12017–12022
25. Gwinn DM, Shackelford DB, Egan DF, Mihaylova MM, Mery A, Vasquez DS, Turk BE, Shaw RJ (2008) AMPK phosphorylation of raptor mediates a metabolic checkpoint. Mol Cell 30:214–226
26. Inoki K, Zhu T, Guan KL (2003) TSC2 mediates cellular energy response to control cell growth and survival. Cell 115:577–590
27. Pinho R, Guedes LC, Soreq L, Lobo PP, Mestre T, Coelho M, Rosa MM, Gonçalves N, Wales P, Mendes T, Gerhardt E, Fahlbusch C, Bonifati V, Bonin M, Miltenberger-Miltényi G, Borovecki F, Soreq H, Ferreira JJ, Outeiro TF (2016) Gene expression differs in peripheral blood of Parkinson's disease patients with distinct progression profiles. PLoS One 12:e0157852
28. Cooper-Knock J, Kirby J, Ferraiuolo L, Heath PR, Rattray M, Shaw PJ (2012) Gene expression profiling in human neurodegenerative disease. Nat Rev Neurol 8:518–530
29. Annesley SJ, Allan CY, Sanislav O, Evans A, Fisher PR (2022) Dysregulated gene expression in lymphoblasts from Parkinson's disease. Proteomes 10:20
30. Houldcroft CJ, Petrova V, Liu JZ, Frampton D, Anderson CA, Gall A, Kellam P

(2014) Host genetic variants and gene expression patterns associated with Epstein-Barr virus copy number in lymphoblastoid cell lines. PLoS One 9(10):e108384

31. Hussain T, Mulherkar R (2012) Lymphoblastoid cell lines: a continuous in vitro source of cells to study carcinogen sensitivity and repair. Int J Mol Cell Med 1(2):75–87
32. Missailidis D, Annesley SJ, Allen CY, Sanislav O, Lidbury BA, Lewis DP, Fisher PR (2020) An isolated complex V inefficiency and dysregulated mitochondrial function in immortalized lymphocytes from ME/CFS patients. Int J Mol Sci 21:1074–1100
33. Annesley SJ, Fisher PR (2021) Lymphoblastoid cell lines as models to study mitochondrial function in neurological disorders. Int J Mol Sci 22(9):4536–4553
34. Annesley SJ, Lay ST, De Piazza SW, Sanislav O, Hammersley E, Allan CY, Francione LM, Bui MQ, Chen ZP, Ngoei KRW, Tassone F, Kemp BE, Storey E, Evans A, Loesch D, Fisher PR (2016) Immortalized Parkinson's disease lymphocytes have enhanced mitochondrial respiratory activity. Dis Model Mech 9(11): 1295–1305
35. Panov AV, Gutekunst CA, Leavitt BR, Hayden MR, Burke JR, Strittmatter WJ, Greenamyre JT (2002) Early mitochondrial calcium defects in Huntington's disease are a direct effect of polyglutamines. Nat Neurosci 5(8):731–736
36. Pansarasa O, Bordoni M, Drufuca L, Diamanti L, Sproviero D, Trotti R, Bernuzzi S, Salvia SL, Gagliardi S, Ceroni M, Cereda C (2018) Lymphoblastoid cell lines as a model to understand amyotrophic lateral sclerosis disease mechanisms. Dis Model Mech 11(3):dmm031625
37. Ugalde CL, Annesley SJ, Gordon SE, Mroczek K, Perugini MA, Lawson VA, Fisher PR, Finkelstein DI, Hill AF (2020) Misfolfded α-synuclein causes hyperactive respiration without functional deficit in live neuroblastoma cells. Dis Model Mech 13(1):dmm040899
38. Chen S, Annesley SJ, Jasim RAF, Musco VJ, Sanislav O, Fisher PR (2017) The Parkinson's disease-associated protein DJ-1 plays a positive nonmitochondrial role in endocytosis in *Dictyostelium* cells. Dis Model Mech 10(10): 1261–1271
39. Haylett W, Swart C, van der Westhuizen F, van Dyk H, van der Merwe L, van der Merwe C, Loos B, Carr J, Kinnear C, Bardien S (2016) Altered mitochondrial respiration and other features of mitochondrial function in *Parkin*-mutant fibroblasts from Parkinson's disease patients. Parkinsons Dis 2016:1819209
40. Miller-Fleming L, Giorgini F, Outeiro TF (2008) Yeast as a model for studying human neurodegenerative disorders. Biotechnol J 3: 325–338
41. Caldwell KA, Willicott CW, Caldwell GA (2020) Modeling neurodegeneration in *Caenorhabditis elegans*. Dis Model Mech 13: dmm046110
42. Sen A, Cox RT (2017) Fly models of human disease: *Drosophila* as a model for understanding human mitochondrial mutations and disease. Curr Top Dev Biol 121:1–27
43. Torija P, Vicente JJ, Rodrigues TB, Robles A, Cerdan S, Sastre L, Calvo RM, Escalante R (2005) Functional genomics in *Dictyostelium*: MidA, a new conserved protein, is required for mitochondrial function and development. J Cell Sci 119:1154–1164
44. Pearce XG, Annesley SJ, Fisher PR (2019) The Dictyostelium model for mitochondrial biology and disease. Int J Dev Biol 63:497–508
45. Francione LM, Annesley SJ, Carilla-Latorre S, Escalante R, Fisher PR (2010) The *Dictyostelium* model for mitochondrial disease. Semin Cell Dev Biol 22:120–130
46. Carilla-Latorre S, Annesley SJ, Muñoz-Braceras S, Fisher PR, Escalante R (2013) Ndufaf5 deficiency in the *Dictyostelium* model: new roles in autophagy and development. Mol Biol Cell 24(10):1519–1528
47. Chida J, Yamaguchi H, Amagai A, Maeda Y (2004) The necessity of mitochondrial genome DNA for normal development of *Dictyostelium* cells. J Cell Sci 117:3141–3152
48. Kotsifas M, Barth C, De Lozanne A, Lay ST, Fisher PR (2002) Chaperonin 60 and mitochondrial disease in *Dictyostelium*. J Muscle Res Cell Motil 23:839–852
49. Carilla-Latorre S, Gallardo ME, Annesley SJ, Calvo-Garrido J, Graña O, Accari SL, Smith PK, Valencia A, Garesse R, Fisher PR, Escalante R (2010) MidA is a putative methyltransferase that is required for mitochondrial complex I function. J Cell Sci 123:1674–1683
50. Wilczynska Z, Barth C, Fisher PR (1997) Mitochondrial mutations impair signal transduction in Dictyostelium discoideum slugs. Biochem Biophys Res Commun 234(1):39–43
51. Francione LM, Fisher PR (2011) Heteroplasmic mitochondrial disease in Dictyostelium discoideum. Biochem Pharmacol 82:1510–1520
52. Storey CL, Williams RSB, Fisher RB, Annesley SJ (2022) *Dictyostelium discoideum*: a model system for neurological disorders. Cell 11:463

53. Mroczek K, Fernando S, Fisher PR, Annesley SJ (2021) Interactions and cytotoxicity of human neurodegeneration-associated proteins tau and α-synuclein in the simple model *Dictyostelium discoideum*. Front Cell Dev Biol 9: 741662
54. McMains VC, Myre M, Kreppel L, Kimmel AR (2010) Dictyostelium possesses highly diverged presenilin/γ-secretase that regulates growth and cell-fate specification and can accurately process human APP: a system for functional studies of the presenilin/γ-secretase complex. Dis Model Mech 3(9–10):581–594
55. Ludtmann MHR, Otto GP, Schilde C, Chen ZH, Allan CY, Brace S, Beesley PW, Kimmel AR, Fisher P, Killick R, Williams RSB (2014) An ancestral non-proteolytic role for presenilin proteins in multicellular development of the social amoeba *Dictyostelium discoideum*. J Cell Sci 127(7):1576–1584
56. Sharma D, Otto G, Warren EC, Beesley P, King JS, Williams RSB (2019) Gamma secretase orthologs are required for lysosomal activity and autophagic degradation in *Dictyostelium discoideum*, independent of PSEN (presenilin) proteolytic function. Autophagy 15:1407–1418
57. Chen S, Annesley SJ, Jasim RAF, Fisher PR (2021) The Parkinson's disease-associated protein DJ-1 protects *Dictyostelium* cells from AMPK-dependent outcomes of oxidative stress. Cell 10(8):1874
58. Chen S, Sanislav O, Annesley SJ, Fisher PR (2018) Mitochondrial HTRA2 plays a positive, protective role in *Dictyostelium discoideum* but is cytotoxic when overexpressed. Genes 9(7): 355
59. Fernando S, Allan CY, Mroczek K, Pearce X, Sanislav O, Fisher PR, Annesley SJ (2020) Cytotoxicity and mitochondrial dysregulation caused by α-Synuclein in *Dictyostelium discoideum*. Cell 9(10):2289
60. Myre MA, Lumsden AL, Thompson MN, Wasco W, MacDonald ME, Gusella JF (2011) Deficiency of huntingtin has pleiotropic effects in the social amoeba *Dictyostelium discoideum*. PLoS Genet 7(4):e1002052
61. Smith PK, Sen MG, Fisher PR, Annesley SJ (2019) Modelling of neuronal ceroid lipofuscinosis type 2 in *Dictyostelium discoideum* suggests that cytopathological outcomes result from altered TOR signalling. Cell 8(5):469
62. Lay S, Sanislav O, Annesley SJ, Fisher PR (2016) Mitochondrial stress tests using seahorse respirometry on intact Dictyostelium discoideum cells. Methods Mol Biol 1407:41–61
63. Presley AD, Fuller KM, Arriaga EA (2003) MitoTracker Green™ labelling of mitochondrial proteins and their subsequent analysis by capillary electrophoresis with laser-induced fluorescence detection. J Chromatogr B Analyt Technol Biomed Life Sci 793(1):141–150
64. Macho A, Decaudin D, Castedo M, Hirsch T, Susin SA, Zamzami N, Kroemer G (1996) Chloromethyl-X-rosamine is an aldehyde-fixable potential-sensitive fluorochrome for the detection of early apoptosis. Cytometry 25:333–340
65. Pendergrass W, Wolf N, Poot M (2004) Efficacy of MitoTracker Green™ and CMXRosamine to measure changes in mitochondrial membrane potentials in living cells and tissues. Cytometry A 61A(2):162–169
66. Rosenkranz AR, Schmaldienst S, Stuhlmeier S, Chen W, Knapp W, Zlabinger GJ (1992) A microplate assay for the detection of oxidative products using 2′,7′-dichlorofluorescin-diacetate. J Immunol Methods 156:39–45

Chapter 2

Behavioral Tests for Associative Learning in *Caenorhabditis elegans*

Aelon Rahmani, Anna McMillen, Ericka Allen, Caitlin Minervini, and Yee Lian Chew

Abstract

Learning is critical for survival as it provides the capacity to adapt to a changing environment. At the molecular and cellular level, learning leads to alterations within neural circuits that include synaptic rewiring, synaptic plasticity, and protein level/gene expression changes. There has been substantial progress in recent years on dissecting how learning and memory is regulated at the molecular and cellular level, including the use of compact invertebrate nervous systems as experimental models. This progress has been facilitated by the establishment of robust behavioral assays that generate a quantifiable readout of the extent to which animals learn and remember. This chapter will focus on protocols of behavioral tests for associative learning using the nematode *Caenorhabditis elegans*, with its unparalleled genetic tractability, compact nervous system of ~300 neurons, high level of conservation with mammalian systems, and amenability to a suite of behavioral tools and analyses. Specifically, we will provide a detailed description of the methods for two behavioral assays that model associative learning, one measuring appetitive olfactory learning and the other assaying aversive gustatory learning.

Key words *Caenorhabditis elegans*, Associative learning, Classical conditioning, Memory, Behavior teKey wordssts, Chemotaxis, Taste/gustation, Olfaction

1 Introduction

1.1 Background

Learning is a vital property for animals to adapt their behavior to a continuously changing environment and is critical for survival. At a cellular level, learning leads to changes within neural circuits. These include rewiring of synaptic connections, alterations in the amounts of presynaptically released neurotransmitters, as well as changes in the number and sensitivity of postsynaptic receptors [1]. Although a remarkable level of understanding has been achieved regarding the molecular changes associated with learning through using

Aelon Rahmani and Anna McMillen contributed equally with all other contributors.

Sebastian Dworkin (ed.), *Neurobiology: Methods and Protocols*, Methods in Molecular Biology, vol. 2746,
https://doi.org/10.1007/978-1-0716-3585-8_2,

invertebrate and vertebrate nervous systems as experimental models, the specific processes that underlie learning and memory in brains of all sizes remain to be revealed. An essential requirement for such research is to establish robust behavioral tests that can measure the extent of learning in these experimental systems.

Two main categories of learning are associative and non-associative learning. Non-associative learning includes habituation and sensitization: habituation is the response decrement as a result of being repeatedly stimulated, whereas sensitization is the response facilitation produced by a novel and/or noxious stimulus [2]. Associative learning refers to processes where a link/association is formed between two cues or behaviors. One type of associative learning is referred to as "classical conditioning," described in more detail in Subheading 1.3. In general, classical conditioning involves pairing a neutral cue with a stimulus that invokes a positive or negative behavioral response. After exposure to both cues simultaneously (pairing or "conditioning"), organisms learn to respond to the neutral stimulus with the same behavioral response as to the stimulatory cue [3]. Classical conditioning is critical for survival as it provides a means for animals to predict the presence of threatening or appetitive cues in their environment.

Model organisms such as *Caenorhabditis elegans* (*C. elegans*), the nematode roundworm, have been used to study short-, intermediate-, and long-term memory formation, as well as both positive (appetitive) and negative (aversive) associative learning using a variety of conditioned stimuli (reviewed in [4]). These include gustatory (taste), olfactory (smell), thermal (temperature), mechanical (touch), and gaseous (e.g., oxygen, carbon dioxide) cues or more complex stimuli that encode a combination of multiple cues, such as the presence of pathogens, food, or mates (reviewed in [5]).

In this chapter, we will focus on behavior tests assaying associative learning in *C. elegans*, specifically assays for *negative association (aversive) gustatory learning* and *positive association (appetitive) olfactory learning*.

1.2 *C. elegans* as a Model System for Learning and Memory Studies

C. elegans has been widely used to study learning and memory, providing significant insight into the cellular and molecular basis of these processes. This is largely due to the organism's amenability to genetic manipulation and well-defined neural circuits. A great strength of this model is that it is the only adult animal with a fully mapped connectome (consisting of ~300 neurons in the hermaphrodite), with sensory neurons and their receptors/modalities having been extensively characterized [6]. Many behavioral paradigms and their relevant neural circuits have been described in the worm, allowing for single-cell investigations and detailed identification of key molecular pathways [5].

The genetic tractability of the worm facilitates the rapid generation of mutant strains and tissue- or cell-specific transgenic lines, allowing researchers to readily probe the role of specific genes and signalling pathways in learning and memory formation [6, 7]. In addition, *C. elegans* husbandry is relatively labor- and cost-effective and does not require extensive facilities. Worms are typically grown on agar plates and fed with a non-pathogenic *Escherichia coli* (*E. coli*) strain that can be cultured in the lab [8]. This, together with the reproductive capacity of *C. elegans* hermaphrodites to generate clonal progeny, enables the rapid generation of large numbers of worms for population-based assays. *C. elegans* strains can also be kept in indefinite, long-term storage at −80 °C, allowing for laboratories to have a large collection of mutant strains that do not require ongoing maintenance. The worm is also amenable to a wide range of techniques for experimental manipulation. For example, the transparent nature of the worm makes it well suited for light-based techniques such as optogenetics and fluorescence imaging, which can be paired with calcium, voltage, and other sensors to probe neural activity in real time in living animals [9, 10]. These tools, in combination with genetic approaches, behavioral tests, and the amenability to high-throughput assays, make *C. elegans* a powerful model to investigate learning and memory at the molecular, cellular, and circuit levels.

1.3 Types of Associative Learning Assays Using C. elegans

C. elegans demonstrates associative learning through several established behavioral paradigms (reviewed in [4, 5]). In *classical conditioning*, animals are presented with an "unconditioned stimulus" and a "conditioned stimulus." An *unconditioned stimulus (US)* will elicit an innate behavioral response called the *unconditioned response*. In contrast, the *conditioned stimulus (CS)* either does not produce an innate response or the innate response differs from the unconditioned response. Following conditioning where the two stimuli are paired together, worms will display the unconditioned response when they are presented with only the conditioned stimulus. This behavioral change is called the *conditioned response*, and it shows that they have learned an association between the US and the CS.

C. elegans behavioral responses are typically measured by their movement toward or away from a stimulus/cue. When worms move away from a cue, this indicates that the cue is repulsive/aversive. Conversely, worms are expected to move toward cues that are attractive/appetitive. For gustatory or olfactory cues, this movement is called *chemotaxis*. The degree of movement toward or away from a cue can be quantified as a *chemotaxis index*, described in Subheadings 3.4.4 and 3.5.3.

This chapter will focus on two behavioral paradigms: "butanone appetitive learning" and "salt aversive learning," which differ in two main ways. Firstly, these paradigms utilize different sensory modalities as conditioned stimuli; *butanone appetitive learning* uses an olfactory cue, and *salt aversive learning* involves a gustatory cue [11, 12]. Secondly, *salt aversive learning* induces an aversive behavioral response toward a conditioned stimulus that was previously appetitive [11]. In comparison, the conditioned stimulus used in *butanone appetitive learning* becomes a strong appetitive cue after training [12]. The demonstrated behaviors after training result from the use of two different unconditioned stimuli, as detailed below.

1.3.1 Butanone Appetitive Learning

Satiation is advantageous to the survival of *C. elegans*, so their natural food source (bacteria) is commonly used as an unconditioned stimulus [12–14]. Previous studies have paired this unconditioned stimulus with the olfactory cue butanone to promote appetitive associative learning, and butanone chemotaxis behaviour can be used for this assay to assess learning [15].

C. elegans naturally have a mild attraction to butanone, but by pairing the odorant as the conditioned stimulus with food, worms learn to develop a stronger attraction to this odorant [12, 15]. This is butanone chemotaxis can be measured using a choice assay where worms can choose to migrate toward butanone or a neutral chemical. This "choice" can be quantified based on this migration through calculating a numerical value called a "chemotaxis index" (Fig. 1). Laboratories that use this behavioral paradigm normally use ethanol (EtOH) as the neutral odorant in these choice assays. This is because butanone is diluted in EtOH when it is presented to the worm during conditioning, so the use of EtOH serves to highlight any behavioral changes introduced by the chemical diluent instead of butanone. Before training/conditioning, the worms show a mild attraction or preference for the butanone, but after conditioning with butanone in the presence of food, they can learn to demonstrate an increased preference for the odorant [12, 15].

C. elegans have two pairs of sensory neurons in the head that are responsible for attraction to specific odorants, AWC and AWA. The neuron pair AWC is solely responsible for the worm's attraction toward the chemical butanone [12, 16]. Unlike many other pairs of neurons in the worm, AWC is asymmetrical, with one of the neurons that expresses the G protein-coupled receptor STR-2 and one that does not [12, 17, 18]. The neuron that expresses STR-2 is stochastically determined and is designated AWC^{ON}, while the other is AWC^{OFF}. Before differentiation, the two AWC neurons interact near each other. This causes repression of calcium signalling and repression of the NSY-1/ASK1 mitogen-activated protein kinase signalling cascade in one of the AWC neurons, which then expresses STR-2 as a result. AWC^{ON} can detect even low-level concentrations of butanone in the worm's environment [12, 18]. An increased attraction to butanone after pairing with food requires the AWC^{ON} neuron and relies on the ciliated ends of

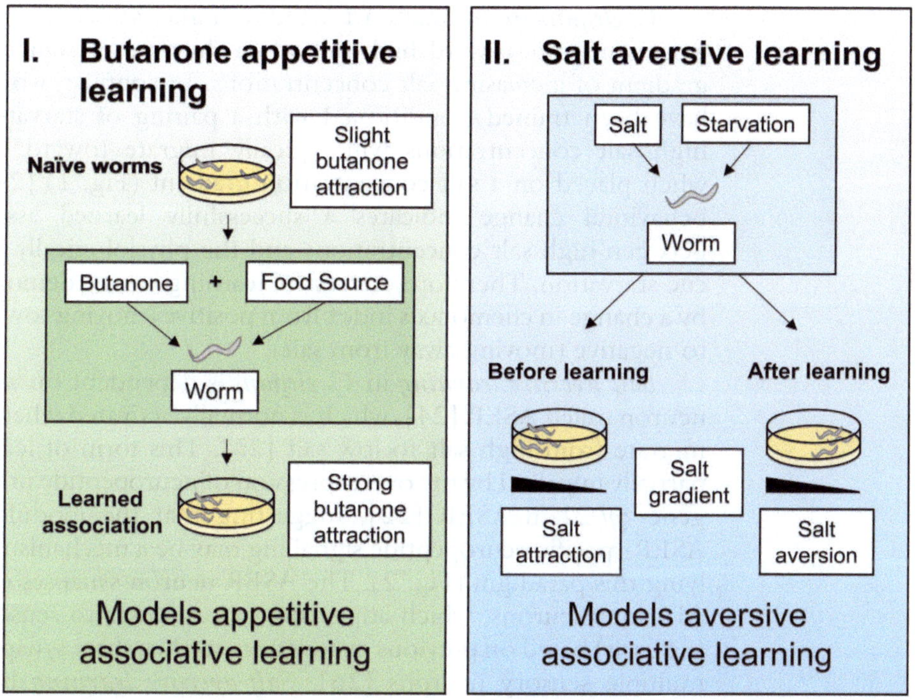

Fig. 1 Schematic of behavioral paradigms in *C. elegans*. This schematic summarizes the two behavioral paradigms outlined in this chapter, namely, *butanone appetitive learning* (left panel) and *salt aversive learning* (right panel). *Left panel*: naïve worms have a slight attraction to the odorant butanone, but when butanone is paired with food during conditioning, the worms demonstrate a stronger attraction to this odorant. *Right panel*: normally, salt is an attractant for the worm, and they will migrate toward high salt on a salt concentration gradient. However, worms can learn to avoid salt through exposure to salt during food depletion or starvation

neurons [19]. The diacylglycerol/protein kinase C signalling pathway can cause repulsion to butanone through the AWC^{ON} neuron rather than attraction. This indicates that AWC^{ON} is important in modifying the worm's level of attraction to butanone both positively and negatively [20]. Although butanone can also be used to study aversive associative learning through pairing the odorant with an aversive stimulus, we will focus on describing its use to study appetitive associative learning in this chapter. Further information on AWC and other sensory neurons sensitive to odorants can be found in reviews on the topic, including [16, 21].

1.3.2 Salt Aversive Learning

In this behavioral paradigm, a no-food (starvation) environment is paired with the gustatory cue sodium chloride (NaCl)/salt [22]. Food depletion or starvation is innately stressful for the worm (reviewed in [23]), so it can function as an unconditioned stimulus. Conversely, salt is the conditioned stimulus for this behavioral paradigm.

C. elegans are normally attracted to higher salt concentrations: worms migrate toward high salt when they are presented with a gradient of increasing salt concentrations. In contrast, worms that have been trained/conditioned with a pairing of starvation and high salt concentrations will typically migrate toward low salt when placed on a salt concentration gradient (Fig. 1) [24]. This behavioral change indicates a successfully learned association between high salt concentrations and the physiologically stressful cue starvation. Therefore, successful learning can be demonstrated by a change in chemotaxis index from positive (moving toward salt) to negative (moving away from salt).

Salt aversive learning in *C. elegans* is dependent on a sensory neuron called ASER [24], which is normally activated when worms migrate from high salt to low salt [25]. This form of learning is partially impaired by the overexpression of neuropeptide processing gene *egl-21* in ASER [22], suggesting that the modulation of ASER-specific neuropeptide signalling may be a mechanism underlying this paradigm (Fig. 2). The ASER neuron synapses onto the AIA interneurons, which adjust motor responses to sensory cues (i.e., salt) based on previous experiences, as they share synapses with multiple sensory neurons [26]. *Salt-aversive learning* has been proposed to occur through a feedback mechanism from AIA to ASER, signalling through an insulin-like peptide called INS-1 [27]. ASER activation is usually correlated with increased levels of a molecule important for neuromodulator release called diacylglycerol (DAG), but this DAG-related response is suppressed after *salt-aversive learning*. ASER has been hypothesized to activate the AIB and AIY bilateral interneurons [28, 29]. AIB is responsible for reorientation, and AIY inhibits turning and reversals [30]. INS-1 likely affects DAG levels [31, 32], so the release of INS-1 from AIA may suppress ASER-specific DAG dynamics to modulate AIB/AIY activation and salt aversion (Fig. 2). For further details on the molecular and cellular mechanisms thought to underlie salt gustatory learning, see detailed reviews [4, 5].

1.4 C. elegans Maintenance for Associative Learning Assays

C. elegans is transparent and can be visualized using a standard dissecting stereomicroscope (10× magnification eyepieces, 0.5–6× magnification objectives) with diascopic base and a transmitted light source. In laboratories, these animals are commonly grown on Petri dishes or "plates" that contain agar, such that they will lay or crawl on top of the agar, with free access to a bacterial food source (e.g., *E. coli* OP50). This agar usually contains a specific composition called "standard nematode growth medium" or NGM [33]. Here, we refer to plates containing this NGM agar as "maintenance plates." Worms grown under laboratory conditions are mostly hermaphrodites that reproduce through self-fertilization, generating clonal progeny. Maintenance of *C. elegans* cultures therefore generally involves transferring hermaphrodite worms

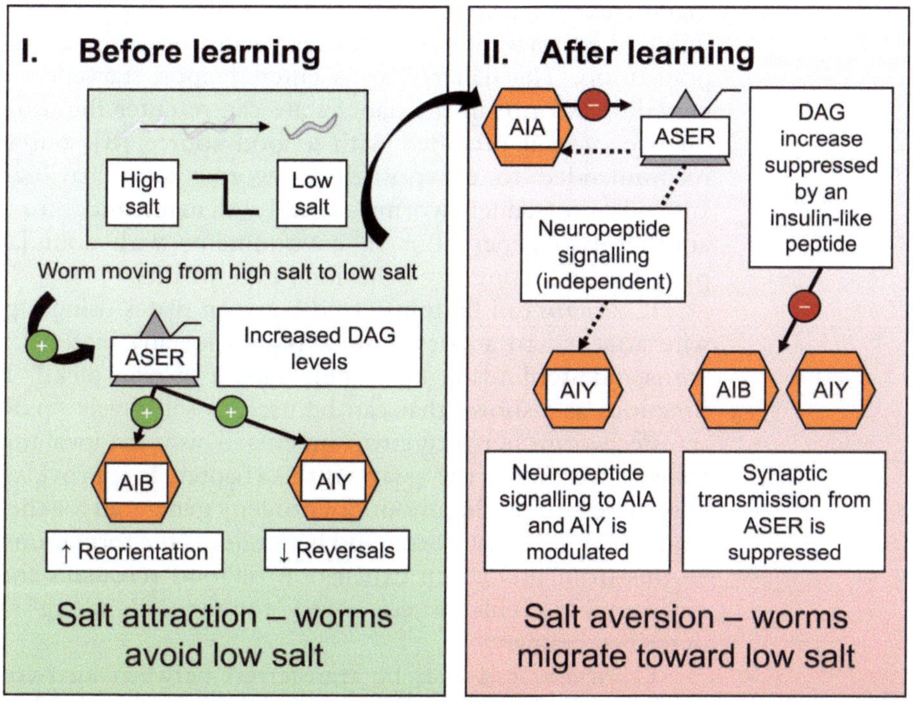

Fig. 2 Schematic of neural circuit for salt aversive learning in *C. elegans*. This schematic displays neural mechanisms before learning (left panel) and after learning (right panel). *Left panel*: AIB/AIY interneurons (orange hexagons) are proposed to become activated by the salt-sensing ASER neuron (gray triangle) upon migration toward low salt (inset panel) [26, 28, 29]. This is likely facilitated by increased diacylglycerol (DAG) levels in ASER [28]. *Right panel*: salt-dependent DAG dynamics may be suppressed by the AIA interneurons via an insulin-like peptide, potentially impacting AIB/AIY [27]. ASER-specific neuropeptide signalling is also proposed as a mechanism for learning [22]. *Symbols*: excitatory neurotransmission = arrows with green circles and a "+" sign; inhibitory neurotransmission = arrows with red circles with a "–"sign; neuropeptidergic signalling = dotted arrows

onto maintenance plates with *E. coli* OP50 to provide sufficient food and space [8, 33]. As each hermaphrodite worm normally generates 100–200 progeny, assuming it does not have a mutation that affects fertility, the maintenance of stocks requires only the transfer of a few (3–4) hermaphrodite worms as required.

C. elegans can be maintained between 15 and 25 °C, with growth rate being proportional to temperature. Worms grown at 15 °C usually take ~7 days to develop from egg to egg-laying hermaphrodite adult, whereas the same process normally takes ~4 days at 20 °C and ~2 days at 25 °C. Different mutant strains can develop at different rates though, so this process must be monitored carefully to ensure that worm strains are age-synchronized upon testing.

Behavioral tests using *C. elegans* are exquisitely sensitive to the animal's physiological state such as age, stress, or nutrient status. It is therefore critical that worms used for behavioral assays are maintained on plates with a sufficient bacterial lawn to prevent unintended starvation and "dauer" state formation in the worm population. The dauer state is entered upon starvation or overcrowding. Worms in the dauer state can re-enter the usual reproductive cycle if provided with a food source [8], but it is not recommended to use post-dauer worms for behavioral assays. Instead, post-dauer worms should be maintained for at least another two generations in environments with abundant food prior to conducting any behavioral experiments.

C. elegans can be transferred between plates using a platinum wire attached to a glass Pasteur pipette or any material that can withstand sterilization by flame, called a "worm pick." The pick functions as a shovel that can be used to selectively "pick" up or transfer worms between environments. This is the ideal method to transfer worms for the assays in this chapter, as it provides greater control regarding the amount of progeny generated for these assays from the transferred worms and hence limits the formation of dauer worms from premature exhaustion of food resources from their environment. This is compared to the "chunking" method described below.

C. elegans can also be transferred between agar-containing plates by cutting out a slice/"chunk" of agar containing a large number of worms and then placing the chunk onto a plate containing abundant food. This "chunking" technique can be used when worms have been starved or have entered the dauer state, to enable two generations to pass in abundant food conditions so that behavioral experiments can be later performed.

This chapter will include step-by-step protocols for "picking" and "chunking" that we use to generate young adult (day 1) hermaphrodite *C. elegans* for associative learning assays.

2 Materials

2.1 Buffers and Solutions

1. Butanone, 10% (v/v): Dilute absolute butanone in 95% (v/v) ethanol (EtOH, detailed below) (*see* **Note 1**).
2. $CaCl_2$, 1 M: 6.95 g/L $CaCl_2$ in Milli-Q water (*see* **Note 2**).
3. Cholesterol, 5 mg/mL: 0.2 g/L ethanol in absolute molecular grade ethanol (*see* **Note 3**).
4. *E. coli* OP50 (*see* **Note 4**).
5. EtOH, 70% (v/v) in Milli-Q.
6. EtOH, 95% (v/v): Dilute absolute molecular grade EtOH in Milli-Q water (*see* **Note 1**).
7. K_3PO_4, 1 M, pH 6.0: 27.10 g/L KH_2PO_4 and 8.90 g/L K_2HPO_4 in Milli-Q water (*see* **Note 2**).

8. LB broth: 1.0% (w/v) tryptone, 1.0% (w/v) NaCl, and 0.5% (w/v) yeast extract in Milli-Q (*see* **Note 2**).
9. M9 buffer: 86 mM NaCl, 42 mM Na_2HPO_4, 22 mM KH_2PO_4, and 1 mM $MgSO_4$ in Milli-Q water (*see* **Notes 2** and **5**).
10. $MgSO_4$, 1 M: 7.53 g/L $MgSO_4$ in Milli-Q water (*see* **Note 2**).
11. NaN_3, 0.1 M: 6.67 g/L NaN_3 in Milli-Q water (*see* **Note 6**).
12. Wash buffer containing salt (i.e., SW buffer): 50 mM NaCl, 5 mM K_3PO_4 (pH 6.0), 1 mM $CaCl_2$, and 1 mM $MgSO_4$ in Milli-Q water (*see* **Note 7**).
13. Wash buffer without salt (i.e., NW buffer): 5 mM K_3PO_4 (pH 6.0), 1 mM $CaCl_2$, and 1 mM $MgSO_4$ in Milli-Q water (*see* **Notes** 7 and **8**).

2.2 Agar Plates

1. Chemotaxis assay (CTX) plates: 2% (w/v) agar, 5 mM K_3PO_4 (pH 6.0), 1 mM $CaCl_2$, and 1 mM $MgSO_4$ in Milli-Q water.
2. "High salt" plates: 2% (w/v) agar, 200 mM NaCl, 5 mM K_3PO_4 (pH 6.0), 1 mM $CaCl_2$, and 1 mM $MgSO_4$ in Milli-Q water.
3. Maintenance plates: 50 mM NaCl, 43 mM K_3PO_4 (pH 6.0), 22 mM $MgSO_4$, 19 mM $CaCl_2$, 13 μM cholesterol in absolute EtOH, 1.7% (w/v) agar, and 0.25% (w/v) tryptone in Milli-Q water.

2.3 Equipment

1. 15 mL tubes with a conical base.
2. Airtight containers.
3. Autoclaved 1.5 mL tubes.
4. Autoclaved glass serological pipette tips.
5. Bottle for liquid waste disposal (*see* **Note 9**).
6. Bunsen burner and/or matches.
7. Forceps.
8. Fume hood.
9. Glass serological pipette tips, 5 mL.
10. KimWipes.
11. Manual P10 pipette, one.
12. Manual P1000 pipettes, two.
13. Manual P200 pipettes, two.
14. Marker pens.
15. Orbital shaker.
16. P10 pipette tips, nonfiltered and autoclaved.
17. P10 pipette tips, filtered.

18. P1000 pipette tips, nonfiltered and autoclaved.
19. P1000 pipette tips, filtered.
20. P200 pipette tips, nonfiltered and low retention.
21. Razor.
22. Serological pipette, one.
23. Stereomicroscope.
24. Sticky tape.
25. Timer.
26. "Worm picks" (*see* Subheading 3.3).

3 Methods

3.1 C. elegans Food Preparation

1. Prepare a streak plate of *E. coli* OP50 from a starter culture (*see* **Note 4**).
2. In a sterile 50 mL conical tube, inoculate 25 mL of LB broth using a single colony from the prepared streak plate, and grow this colony on an operating orbital shaker set to 37 °C and 200 rpm for 16–18 h. Liquid culture can be stored at 4 °C for months before use.
3. For ongoing use, liquid *E. coli* culture can be prepared by using 5 μL of *E. coli* OP50 to inoculate 25 mL of LB broth and then incubation of this culture at 37 °C, 200 rpm for 16–18 h.
4. We recommend pipetting a small volume of the *E. coli* liquid culture onto maintenance plates (*see* Subheading 3.2) to check for contamination prior to preparing large numbers of plates. *E. coli* OP50 bacterial lawns should appear homogeneous, light beige in color, with a "shiny" surface.

3.2 Pouring Agar Plates

1. Autoclave Milli-Q water containing agar, NaCl, and tryptone as specified in Subheading 2.2 for the type of plates being made, and then cool to 55 °C. Sodium chloride is only added to maintenance plates at this time, not "High salt" plates.
2. In sterile conditions add 1 M K_3PO_4, 1 M $CaCl_2$, 1 M $MgSO_4$, and 5 mg/mL cholesterol in absolute EtOH as specified in Subheading 2.2 for the type of plates being made. Add NaCl to "High salt" plates at this time as well. Swirl this bottle to mix and then aliquot 10 mL of agar into sterile Petri dishes with 60 mm diameter. A peristaltic pump ("plate pourer") can be used for this purpose—this is recommended over manually aliquoting 10 mL volumes by pipetting, as a pump will reduce the risk of repetitive strain injury when preparing hundreds of agar plates (often required for large-scale maintenance and behavioral assays).

3. Allow the agar to dry overnight at room temperature; the agar will appear slightly translucent and beige/yellow in color when sufficiently dried (*see* **Note 10**).
4. For maintenance plates only, pipette 100 μL of *E. coli* OP50 liquid culture onto the center of each maintenance plate in sterile conditions. Gently swirl the plates to spread the bacteria, without the bacteria reaching the plate edge—this prevents worms from crawling too close to the edge of the agar plate; worms that crawl up the side of the Petri dish rapidly desiccate. Allow the bacteria on these plates to dry overnight at room temperature (*see* **Note 10**).
5. Plates are stored agar side-down as condensation can form on the Petri dish lid.

3.3 C. elegans Maintenance

3.3.1 Picking

1. Use a Bunsen burner or matches to sterilize the platinum wire component of your pick before transferring any worms (*see* **Note 11**). Wait a few seconds before picking so that the wire does not burn any worm you intend to transfer. We recommend that you perform this sterilization process between picking worms with different genotypes as well.
2. Use the pick and a stereomicroscope to physically "pick up" a young adult hermaphrodite worm and transfer the animal onto a new maintenance plate (*see* **Note 12**).
3. Close the lid on your maintenance plates and incubate them agar side-up at 22 °C for approximately three to four days, not including the day that you picked your worms onto the maintenance plates (*see* **Note 13**). The worms you transfer will generate progeny that will constitute your experimental groups.

3.3.2 Chunking

1. Using a sterilized tool that can cut a clean line through NGM media (i.e., a pipette tip or a sterilized blade), cut out a chunk of NGM that has plenty of worms on the surface (*see* **Note 14**).
2. Transfer the chunk onto a fresh maintenance plate that has been inoculated with *E. coli* OP50, so that the side occupied by worms is in direct contact with the agar of the new maintenance plate next to the dried OP50 patch. This transfer can be done by either balancing the square chunk on top of the sterilized tool and "flipping" the square onto the new plate or by using sterilized tweezers to transfer the square.
3. Allow some time for the worms to crawl onto the new plate, before storing the plate agar side-up.

3.4 Butanone Appetitive Learning Assay

Four cohorts are required for each genotype tested: "naïve," "prestarved," "mock-conditioned," and "conditioned" (Fig. 3) (*see* **Notes 15** and **16**).

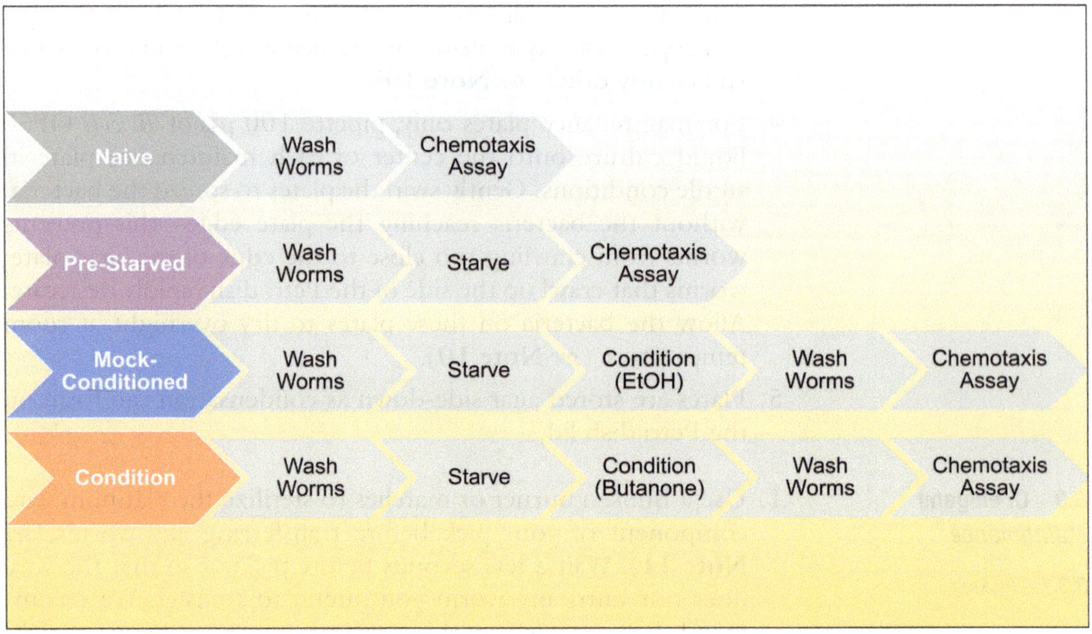

Fig. 3 Workflow for each cohort in butanone appetitive learning assay. For every genotype assayed, there will be four cohorts: the "naïve" cohort that does not undergo conditioning, the "pre-starved" cohort that does not undergo conditioning but is starved prior to testing, the "mock-conditioned" cohort that undergoes conditioning before testing but with 95% (v/v) EtOH that is used as the diluent for butanone, and the "conditioned" cohort that is trained with 10% (v/v) butanone diluted in 95% (v/v) EtOH prior to testing. This figure shows the recommended workflow for each cohort, where "wash worms" refer to Subheading 3.4.1, "starve" refers to Subheading 3.4.2, "condition" refers to Subheading 3.4.3, with chemical in parentheses indicating EtOH (ethanol, for mock-conditioned cohort) or butanone (for conditioned cohort), and "chemotaxis assay" refers to Subheading 3.4.4

3.4.1 Transferring Worms from Agar Plates (The "Washing" Step)

1. Line up all the maintenance plates for each cohort and then pipette 2 mL of fresh M9 buffer onto the furthermost maintenance plate (*see* **Note 16**).
2. After swirling the plate to dislodge worms from bacteria, pour the M9 buffer from this plate to the next plate down the line.
3. Repeat **step 2** until all the plates have had the M9 buffer added and swirled. Once the M9 buffer with the suspended worms reaches the furthest plate on the opposite side, repeat **step 2** in the opposite direction such that each plate is swirled twice. To do this, swirl your current plate containing M9 buffer and worms, and then pour the buffer containing worms onto the nearest plate in line, repeating for all maintenance plates until your worm-containing buffer is back onto the first maintenance plate that had been washed.
4. Pipette all worms suspended in the M9 buffer from the first maintenance plate into a 1.5 mL tube and then incubate worms in that tube for 2 min (*see* **Notes 17** and **18**).

5. Discard supernatant M9 buffer from the tube, trying to limit disturbance of the worm pellet as much as possible (*see* **Note 9**).
6. Pipette 1 mL of fresh M9 buffer onto the worm pellet and incubate for 1 min.
7. Discard supernatant M9 buffer from the tube, trying to limit disturbance of the worm pellet as much as possible. There should be around 100–200 μL of worm-containing solution left in the tube (*see* **Note 9**).
8. The "naïve" cohort will proceed directly to the assay in Subheading 3.4.4, whereas the three remaining cohorts will proceed to Subheading 3.4.2.

3.4.2 Starving Worms in Solution

1. Pipette the entire worm pellet into a new 1.5 mL microcentrifuge tube with a low retention P200 tip, and then add fresh M9 buffer up to the 1 mL mark (*see* **Note 19**).
2. Tape the 1.5 mL tube from **step 1** onto its side on the surface of an orbital shaker. Incubate the worms on this shaker at room temperature for 1 h while it is set at 175 rpm.
3. Remove the tube and sit it upright for 2 min at room temperature to allow worms to pellet again.
4. Discard supernatant M9 buffer from the tube, trying to limit disturbance of the worm pellet as much as possible (*see* **Note 9**). There should be around 100–200 μL of worm-containing solution remaining in the tube.
5. The "pre-starved" cohort will move directly from this procedure to the assay outlined in Subheading 3.4.4. The other two cohorts ("mock-conditioned" and "conditioned") will proceed to Subheading 3.4.3.

3.4.3 Conditioning Worms on Plates

1. Prepare maintenance plates as instructed in Subheading 3.2. Before the assay is to occur, pour these plates 6 days in advance, and add liquid *E. coli* OP50 (i.e., "food") one day in advance (*see* **Note 20**). Label all plates with the genotype and cohort that the plate will be used for prior to beginning the assay (*see* **Notes 21** and **22**).
2. Pipette 33 μL of worms following the starvation protocol in Subheading 3.4.2 next to food onto the respective maintenance plates with a low retention P200 tip (*see* **Notes 19, 23**, and **24**). Dab off excess M9 buffer from the maintenance plates using a clean KimWipe (*see* **Notes 12** and **25**).
3. Use a KimWipe to remove any condensation from the lids of your maintenance plates.

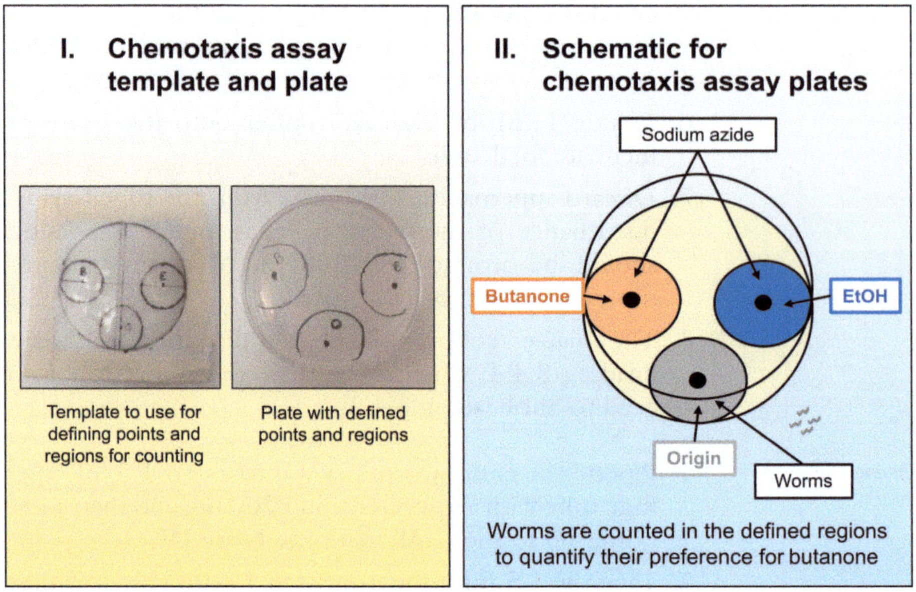

Fig. 4 Schematic for the CTX plates used in the butanone appetitive learning assay. Three dots are drawn, labelled "B" for butanone, "E" for ethanol, and "O" for origin (the point at which the worms will be added). Left panel: shows the template and a plate with the labelled dots and regions drawn. The template contains the circles of the regions for each dot, which when lined up with the dots drawn on the plate can serve as a guide for counting (*see* **Note 2.5**). Right panel: shows a schematic for what the regions represent and what will be added to each dot during the Subheading 3.4.4. Chemotaxis Assay procedure

4. Pipette 6 μL of odorant in a streak on the lid of each maintenance plate. The "mock-conditioned" cohort will use the 95% (v/v) EtOH as the odorant and the "conditioned" cohort will use the 10% (v/v) butanone.
5. Replace lids on maintenance plates and put them into an airtight container for 1 h.
6. Following the one-hour incubation, repeat the washing steps outlined in Subheading 3.4.1, before proceeding to Subheading 3.4.4.

3.4.4 Butanone Chemotaxis Assays

1. Prior to starting the assay, label all CTX plates with three points "B," "E," and "O" as shown in Fig. 4 (*see* **Note 22**). Also include the genotype and cohort that the plate will be used for (*see* **Note 21**). It may be helpful to create a template like shown in Fig. 4 ahead of time to be used throughout all occurrences of the butanone assay to keep counting consistent.
2. For the "pre-starved" cohort, add 1 μL of 0.1 M NaN_3 to "E" and "B" points on each CTX plate 4 min before completion of the 1 h starvation period outlined in **step 2** of Subheading 3.4.2. For all other cohorts, the NaN_3 should be added to CTX plates before the experimenter proceeds with the protocol

outlined in Subheading 3.4.1. The time to do this for "mock-conditioned" and "conditioned" cohorts is during the washing step after conditioning.

3. Add 1 μL of 10% (v/v) butanone to "B" and 1 μL of 95% (v/v) EtOH to "E" on CTX plates (*see* **Note 18**).
4. Pipette 33 μL of worms onto "O" point on CTX plates with a low retention P200 tip (*see* **Notes 19**, **23**, and **24**). Dab off excess M9 buffer from the CTX plates using a KimWipe (*see* **Notes 12** and **25**).
5. Put the CTX plates in an airtight container for 1.5 h, and then move to a 4 °C fridge for 1 h (*see* **Note 26**).
6. Use a plate template to quantify the results of the chemotaxis assay (Fig. 4). Calculate the chemotaxis index (CI) for each plate using the following formula:

$$\mathrm{CI} = \frac{\text{Worms on 'butanone region'} - \text{Worms on 'EtOH region'}}{\text{Total worms} - \text{Worms on 'origin region'}}$$

A CI of zero indicates no preference, while a positive CI indicates preference of butanone, and a negative CI indicates a preference for ethanol. The closer the CI is to ±1, the more of a preference there is. Experimenters should strive to have a total number of worms between 50 and 250 for each technical replicate.

7. Figure 5 shows example data that was generated by following the procedures detailed above.

3.5 Salt Aversive Learning Assay

Experiments that aim to study *salt aversive learning* will require young (day 1) adult hermaphrodite *C. elegans* prepared as described in Subheading 3.3.1 (*see* **Note 15**).

3.5.1 Transferring Worms from Agar Plates (The "Washing" Step)

1. Prepare one serological pipette (*see* **Note 17**), and clean glass serological pipette tips with a 5 mL capacity, two P1000 pipettes set to "1000 μL" with unfiltered tips, one P200 pipette set to "100 μL" with a low retention unfiltered tip, a timer set to 2 min, and a bottle for liquid waste (*see* **Note 9**).
2. Place 4–6 maintenance plates containing worms in a line on a laboratory bench to constitute one experimental group for one genotype. These plates will need to be placed agar side-down without the lids.
3. Tilt a maintenance plate containing worms at a 45° angle facing toward you, and then use a glass serological pipette tip to dispense 4 mL of SW buffer onto the topmost point of the agar. The worms will begin to dislodge from the dried bacterial patch on the maintenance plate.

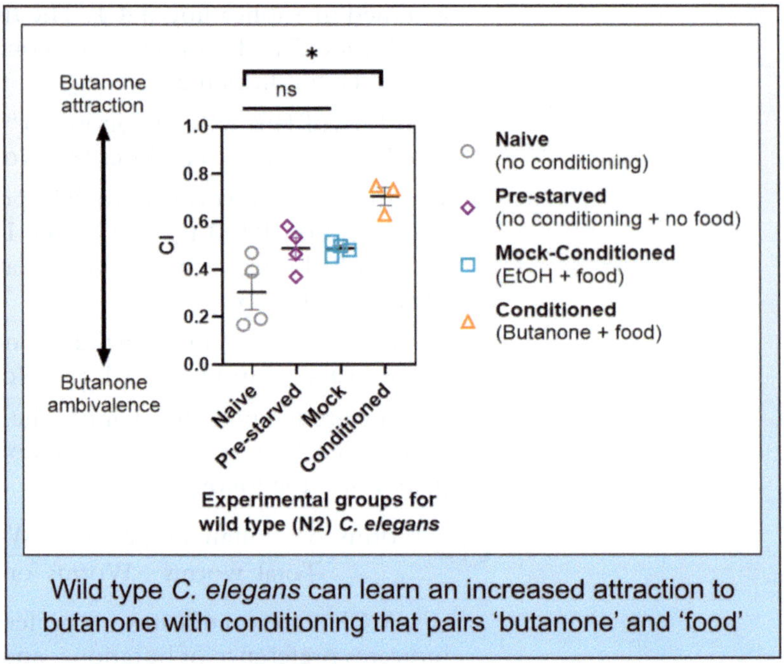

Fig. 5 Example chemotaxis assay results for the butanone appetitive learning assay. The *C. elegans* strain N2, as shown in this graph, has been previously used as the wild-type (control) strain in similar assays as they have been shown to learn normally [12]. Worms in each region of the CTX plates as shown in Fig. 4 were counted and the chemotaxis index (CI) was calculated as described in Subheading 3.4.4. Chemotaxis Assay. CTX plates that contained <50 worms or >250 worms were not included. Each data point represents one biological replicate ($n = 4$) that was generated with four technical replicates (i.e., four CTX plates) per cohort. Data points on the graph represent the mean for each biological replicate. Error bars signify $\pm$ SEM. A one-way ANOVA and Tukey's multiple comparison test was performed (* ≤ 0.05 and ns = no significance). *Symbols*: gray circle = naïve; purple diamonds = pre-starved; blue square = mock-conditioned; orange triangle = conditioned

4. While you are still tilting the maintenance plate, place the same serological pipette tip on the bottommost segment of the plate, and then collect the solution containing worms (*see* **Note 27**).
5. Re-dispense the SW buffer containing worms from **step 4** onto the topmost point of the next maintenance plate in line while it is tilted at a 45° angle.
6. Repeat **steps 4** and **5** until all 4–6 maintenance plates are washed, and then, gently dispense the worm-containing SW buffer in a 15 mL tube (*see* **Notes 28** and **29**).
7. As this assay involves a starvation (no food) conditioning step, worms will need to be washed to minimize the presence of food. The presence of the bacterial food source can be visibly

seen, as washing buffers that contain food will appear cloudy. For the first wash, use the timer to wait <2 min so that worms can settle by gravity as a pellet, and then remove the supernatant as liquid waste with a P1000 without disturbing the worm pellet (*see* **Note 9**).

8. For the second wash, use the remaining P1000 to add 1 mL of SW buffer to the worm pellet, and allow <1 min for the worm pellet to form again. The same P1000 can be used to discard the supernatant, without disturbing the worm pellet, excluding approximately 100 μL of remaining solution containing worms (*see* **Note 9**).
9. For the "naïve" cohort, proceed immediately with the protocol outlined in Subheading 3.5.3.
10. Worms for the "mock-conditioned" and "conditioned" groups will need to be washed a third time. Repeat **step 7** with SW buffer for "conditioned" groups and NW buffer for "mock-conditioned" groups. For each cohort, leave approximately 100 μL of solution including your entire worm pellet in the 15 mL tube and then proceed to Subheading 3.5.2.

3.5.2 Conditioning Worms in Solution

1. Use your prepared P200 with a low retention unfiltered tip to transfer all 100 μL of the solution containing worms from the washing procedure in Subheading 3.5.1 to a 1.5 mL tube (*see* **Notes 19** and **29**).
2. Quickly add the same buffer you used to perform your final wash to your worm pellet so that the final volume of your worm-containing solution is 1 mL (i.e., SW buffer for "conditioned" groups and NW buffer for "mock-conditioned" groups). Swirl the 1.5 mL tube so that worms do not pellet, and then sticky tape the tube on its side to an orbital shaker.
3. Set the orbital shaker to 175 rpm and shake the solutions containing worms at room temperature for 3 h.
4. Remove 1.5 mL tubes from the orbital shaker and sit these tubes upright for <2 min to pellet your worms. Remove the supernatant as liquid waste until approximately 100 μL of solution including your entire worm pellet remains, and then proceed with Subheading 3.5.3 (*see* **Note 9**).

3.5.3 Salt Chemotaxis Assays

1. Prepare each "high salt" plate and CTX plate in 60 mm diameter plates as described in Subheading 3.2, excluding the addition of bacteria. "High salt" plates should have the same composition as CTX plates but also an additional 200 mM NaCl, so that both types of agar plates can be later used to generate salt concentration gradients on CTX plates. These plates can be used immediately once the agar is dry (*see* **Notes 30** and **31**).

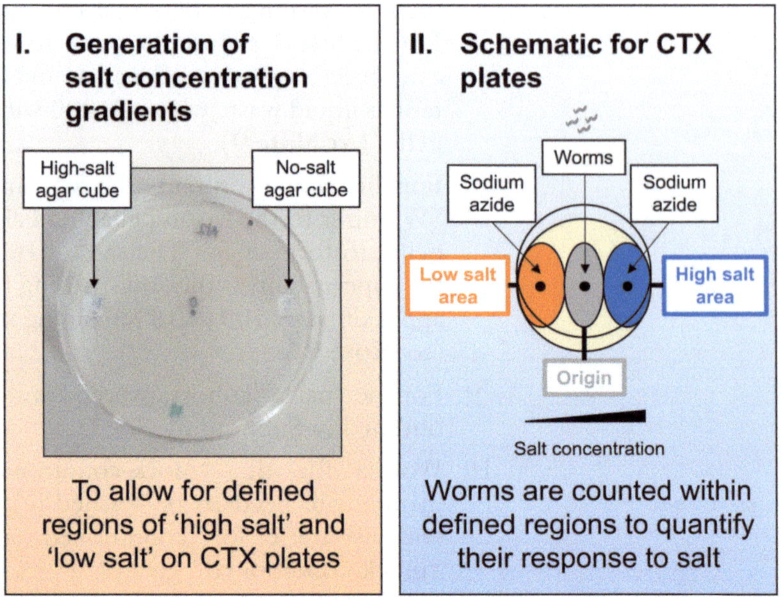

Fig. 6 Schematic for the CTX plates used in the salt aversive learning assay. The CTX plates displayed in the two above schematics can be used to assay salt aversive learning. Left panel: a salt concentration gradient (0–200 mM NaCl) can be generated by overnight incubation of 5 mm cubes of agar with 200 mM NaCl (salt) or without salt on CTX plates. These cubes should be placed approximately 5 mm from the CTX plate edge on positions marked as "H" and "L," respectively. Right panel: regions of high salt (orange) and low salt (blue) can be defined as an ellipse (short axis = 3 cm; long axis = 4 cm). Worms are pipetted onto the region of origin (green), which consists of a 2 cm short axis and 4 cm long axis, to commence chemotaxis assays. Points where agar cubes or worms are transferred onto are represented by a small black circle. Sodium azide is used to paralyze animals when they crawl onto a "high salt" or "low salt" region, to capture their initial behavioral response toward salt. The remaining unlabelled regions in yellow are referred to as "undefined regions"

2. Prepare a "salt CTX plate template" to draw your points of "high salt" (H), "origin" (O), and "low salt" (L) on the same positions of each CTX plate 1 day before the behavioral experiment. To do this, draw onto a thin sheet of plastic with a marker using the exact measurements described in Fig. 6, similar to the template in Fig. 4. In short, "H" and "L" should be 5 mm away from the plate edge and equidistant from each other, whereas "O" should be drawn on the center of each 60 mm diameter CTX plate.
3. Use your "salt CTX plate template" to label each CTX plate you plan to use with three points, the letters "H," "O," and "L," and abbreviations for the strain/s you will use as in Fig. 6 (*see* **Notes 21** and **30**).

4. Place each CTX plate agar side-down with the lid beside it. Carefully sterilize a razor and forceps with 70% (v/v) ethanol using a KimWipe. Use the clean razor to cut 5 mm × 5 mm cubes/plugs from a spare CTX plate and place these plugs onto the dot below "L" for each plate using forceps.
5. Repeat **step 4** but use "high salt" plates and place the plugs onto the dot below "H" for each CTX plate (*see* **Note 32**).
6. Use a KimWipe to thoroughly remove any condensation on the CTX plate lids (*see* **Note 33**), and then place these dried lids onto their respective CTX plates. Store these plates agar side-down in a suitable container in an incubator set to 22 °C for 16–24 h (*see* **Note 34**), being careful to not disturb the plugs that you have placed on your CTX plates.
7. Use a clean KimWipe to dry any condensation from the lids of these plates again before use (*see* **Note 33**).
8. From each CTX plate, remove the plugs used to generate your salt gradients <2 min before use. Pipette 1 μL of 0.1 M NaN_3 to the "L" and "H" points on each CTX plate <1 min before you use the CTX plates (Fig. 6).
9. Pipette 33 μL of your worm-containing solution with a P200 and a low retention unfiltered tip onto the "O" point of each CTX plate according to Fig. 6 (*see* **Notes 19**, **23**, and **24**).
10. Quickly fold a KimWipe in half five times tightly, and then, press the corner of the folded KimWipe onto the middle of each droplet to remove excess solution. Gently press onto the area a maximum of three times to remove residual solution from your worms (*see* **Note 12**). This KimWipe can also be used to wipe any condensation on CTX plate lids, if needed (*see* **Note 33**).
11. Incubate the CTX plates from **step 10** agar side-down in an incubator set to 20–22 °C for 40 min, to allow worms to migrate to their preferred region on the salt concentration gradient. Worms that migrate near "L" and "H" points of CTX plates will become paralyzed by the sodium azide (Fig. 6); this captures their initial chemotaxis behavior in response to salt.
12. Transfer CTX plates to a fridge to immobilize any worms that have not been paralyzed since their final positions are in regions outside of "high salt," "low salt," and "origin" regions (i.e., "undefined regions") (Fig. **6**).
13. For each CTX plate, use a preprepared "salt CTX plate template" to count the number of worms in the undefined regions and in regions of "high salt," "origin," and "low salt" under a stereomicroscope (Fig. 6). These values are used in the following formula to calculate the "chemotaxis index" (CI):

$$CI = \frac{\text{Worms on 'high salt region'} - \text{Worms on 'low salt region'}}{\text{Total worms} - \text{Worms on 'origin region'}}$$

CIs represent how an entire population of *C. elegans* responds to salt. The lowest value for a CI is −1.0, and this would indicate complete salt aversion displayed by all worms in a population. Conversely, complete salt attraction in a worm population would be represented by a CI of +1.0. Each CTX plate generates one CI value (one technical replicate). The mean CI value from all technical replicates of the same experimental group in a biological replicate can then be used to infer how animals within that group will generally respond to salt.

14. Figure 7 shows example data for this learning assay.

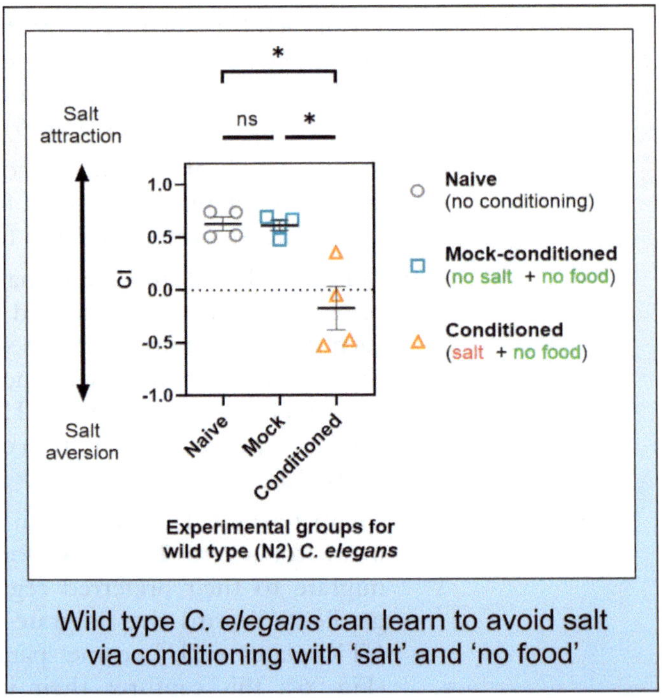

Fig. 7 Example chemotaxis assay results for the salt-aversive learning assay. The *C. elegans* strain N2, as shown in this graph, has been previously used as the wild-type (control) strain in similar assays as they have been shown to learn normally [11]. Chemotaxis index (CI) = (#high salt − #low salt)/(#total − #origin) (Fig. 6). Each data point is one biological replicate ($n = 4$), each with three technical replicates. 81–914 worms were counted per technical replicate. Error bars signify mean ± SEM. Statistical analyses were performed by one-way ANOVA and Tukey's multiple comparisons test (* ≤0.05; ns = no significance). *Symbols*: gray circle = naïve; blue square = mock-conditioned; orange triangle = conditioned

4 Notes

1. 95% (v/v) EtOH and 10% (v/v) butanone needs to be prepared fresh on the day of the *butanone appetitive learning* assay. For each genotype, you will need 35 μL of 10% (v/v) butanone and 35 μL of 95% (v/v) EtOH during this assay, plus the 95% (v/v) EtOH required to dilute butanone. All work with butanone needs to occur in a fume hood. Perform all safety procedures required to safely handle butanone, including wearing two layers of gloves to avoid skin contact with this chemical. Working solutions for EtOH and butanone can be stored at room temperature until ready to use in *butanone appetitive learning* assays.
2. This solution can be generated by autoclaving the respective powder/s in Milli-Q water and then allowing it to cool to room temperature before use. The solution can be stored for months at room temperature. Do not allow contaminants to be introduced to the solution by only opening it in sterile conditions. Make sure to swirl the buffer before use.
3. We recommend that cholesterol stock solutions are prepared in an autoclaved bottle that has been cooled previously and, before use, the bottle has remained closed when in non-sterile conditions.
4. A starter culture can be sourced directly from the Caenorhabditis Genetics Center (CGC) to generate liquid cultures of OP50. Use aseptic technique in all aspects of *E. coli* OP50 liquid culture preparation.
5. M9 buffer is commonly used in studies with *C. elegans* as it is non-toxic to the worms and often used as a neutral solution [8]. This buffer needs to be prepared prior to the start of the *butanone appetitive assay.*
6. NaN_3 solutions need to be prepared prior to the start of both assays. It can be stored for months at 4 °C, so it can be made up at larger quantities ahead of time. For each genotype, 35 μL of this solution is used in *butanone appetitive learning*, and 18 μL is required for *salt aversive learning* for a single biological replicate of one genotype.
7. We recommend using autoclaved Milli-Q water to dilute concentrated buffers in sterile conditions to generate this solution. These buffers can be stored at room temperature and can be used for several months so long as you take care to not introduce contaminants to the solution. Make sure to swirl the buffer before use.
8. Since we measure *salt aversive learning* through a behavioral change, we need to ensure that this change is due to learning and not purely because of exposure to starvation. To do this,

we train a second worm population through starvation in the absence of salt and then assay their salt chemotaxis. This "mock-conditioned" population functions as a negative control since they usually remain attracted to salt, similar to untrained worms [11], confirming that salt aversion is induced by specifically pairing salt and starvation. We generate this negative control using an "NW buffer," which has the same composition as the SW buffer excluding NaCl.

9. Liquid waste should be discarded in a bottle that is able to safely withstand chemical disinfection.
10. Plates should be used within 1–2 months independent of bacteria preparation and can be stored at 4 °C or at room temperature, bearing in mind that the agar plates will dry out (i.e., or expire) faster at higher temperatures.
11. Picks can be made by gently inserting a small length of platinum wire into a hollow glass rod and then carefully melting the glass under a Bunsen burner so that the wire is secured to the rod. A flat, solid tool can then be used to flatten the end of the wire. The flattened end of the worm pick in a "scoop/shovel" shape makes it easier to pick individual worms during each transfer.
12. It is recommended to avoid poking holes in the agar as much as possible since *C. elegans* tend to burrow into these spaces and will no longer be accessible on the agar surface.
13. This number of days will change depending on the genotype of the worm. We recommend monitoring the growth rate of your worms at 22 °C daily before experimentation.
14. The size of the chunk will depend on the state of the original plate. If the plate does not have many worms, it would be recommended to cut a larger chunk. Be aware that chunking a large number of worms onto a plate may cause that particular worm population to starve before the next planned maintenance.
15. These assays can be performed in a single day, and, during this day, each experimental group can be prepared separately in staggered blocks of time with experience. This staggered preparation reduces the risk of mixing genotypes or experimental groups. For *butanone appetitive learning*, all cohorts excluding the "naïve" cohort can be prepared sequentially for a single genotype. The "naïve" cohort can be prepared at this time or during the one-hour conditioning step. For *salt aversive learning*, we normally start by preparing the "mock-conditioned" and "conditioned" groups at the same time for one *C. elegans* genotype. During the three-hour conditioning step for these groups, we then prepare the "naïve" cohort for the same genotype, since the "naïve" cohort does not undergo conditioning.

16. We prepare three maintenance plates for each genotype tested in the *butanone appetitive learning* assay. To cultivate enough worms for this assay, we normally transfer six L4 hermaphrodites per maintenance plate (per genotype), 6 days ahead of the assay. You may need to prepare more or fewer maintenance plates depending on the total number of worms obtained following the chemotaxis assay procedure, as it is important to have enough worms to qualify population-based behavior without overcrowding.
17. We normally wash worms with (A) a serological pipette and sterile, autoclaved glass serological pipette tips or with (B) a P1000 pipette and filtered P1000 pipette tips, depending on preference. In our experience, worms seem less likely to stick onto the sides of glass tips, compared to P1000 tips. Therefore, worm loss can be potentially minimized by using glass tips. Overall waste can also be reduced through the use of glass tips since they can be cleaned and reused for future experiments.
18. Addition of butanone/EtOH can be performed for *butanone appetitive learning* during **step 4** (2 min incubation) in the washing protocol in Subheading 3.4.1 that directly precedes Subheading 3.4.4 procedures for the "naïve," "mock-conditioned," and "conditioned" cohorts or during the 2 min incubation during **step 3** in the starvation protocol in Subheading 3.4.2 for the "pre-starved" cohort.
19. The use of low retention P200 tips can minimize loss of worms when you transfer them during your experiment, compared to standard pipette tips.
20. For *butanone appetitive learning*, there needs to be enough plates made for the "conditioned" and "mock-conditioned" cohorts of each genotype so that three plates can be designated to each cohort during the conditioning step.
21. We suggest labelling worm plates for each cohort with different colors so that it is easier on the day of the experiment to distinguish the plates you need for each *C. elegans* strain and experimental group you will generate.
22. We recommend labelling all plates and tubes for *butanone appetitive learning* before the assay. Alternatively, this can occur when all relevant experimental groups undergo starvation, if needed.
23. When worms are aliquoted to commence chemotaxis assays, if there is an excess of buffer containing worms in the tube afterward, divide this excess buffer as evenly as possible between the plates.
24. Try to pipette a similar number of worms suspended in the buffer onto all sample plates. Slightly agitating the tube to resuspend the worm pellet in the buffer helps with this.

25. Do not fold the KimWipe too tightly and dab lightly once to remove excess liquid without losing many worms.
26. For *butanone appetitive learning*, worms that migrate near "B" and "E" points of CTX plates will become paralyzed by the sodium azide (Fig. 4), capturing their initial preference for butanone or EtOH diluent. After 1.5 h, the refrigeration of these plates at 4 °C renders worms immobilized at their final position during the assay if they do not migrate near "B" or "E" points.
27. Make sure not to introduce bubbles during the pipetting process, as this can provide unnecessary mechanosensation that may stress the animals, negatively impacting their locomotion capacity. Formation of bubbles/aerosols can be minimized by ensuring that the end of the pipette tip is submerged in the solution without pressing onto the agar.
28. The process of transferring worms in liquid and then pelleting them by gravity must be performed as efficiently as possible, as crowding the worms in a pellet for too long (>2 min) can induce an aversion to salt that is not dependent on conditioning/training status.
29. It is critical to pipette solutions containing worms onto the bottom and not the sides of the tube, so that worms are not stuck on the sides of these tubes, to maximize the final number of worms used in your chemotaxis assays.
30. The number of CTX plates required per biological replicate depends on the number of genotypes assayed and the number of technical replicates. Assays for *salt aversive learning* involve three distinct conditions: the "conditioned" group, the "mock-conditioned" group, and the untrained/"naïve" group. We recommend performing at least three chemotaxis assays for each group (i.e., three biological replicates), per genotype assessed, and for each biological replicate to involve at least three CTX plates per group (i.e., three technical replicates). The CTX plates required for the salt chemotaxis assay can be usually prepared with 1–2 "high salt" plates.
31. For the *salt aversive learning* assay, make sure to set aside one extra CTX plate. This plate can be used to generate "no salt" plugs when you set up your salt concentration gradients as described in Subheading 3.5.3.
32. In the instance that a "high salt" plug is accidently dropped onto a site other than at "H," you must discard the plate, since the accidental exposure will compromise your salt concentration gradient for *salt aversive learning*.
33. This is critical for *salt aversive learning* since condensation can disturb the generation of your salt concentration gradients by dripping water onto the agar of your CTX plates during overnight incubation.

34. Salt concentration gradients for *salt aversive learning* could theoretically be formed at other temperatures, but the duration required to form these gradients may vary. It is important though that the agar in these plates is at a temperature that is not lethal for the worms. We have chosen approximately room temperature (i.e., 22 °C), since worms can live in temperatures between 15 and 25 °C [8], and the worms move faster when incubated at higher temperatures.

Acknowledgments

The authors gratefully acknowledge the *Caenorhabditis* Genetics Center, which is supported by the National Institutes of Health (P40 OD010440), for providing strains used in data presented here. We also acknowledge our funding sources: Y.L.C is funded by the National Health and Medical Research Council (NHMRC) (GNT1173448), the Rebecca L Cooper Medical Research Foundation (PG2020652), the Flinders University Impact Seed Funding Grant (2022), and the Flinders Foundation (Mary Overton Senior Research Fellowship). A.R is funded by a PhD scholarship supported through the Australian Graduate Research Training Program. A.L.M. is supported by a Flinders University Research Scholarship (Flinders University).

References

1. Magee JC, Grienberger C (2020) Synaptic plasticity forms and functions. Annu Rev Neurosci 43:95–117. https://doi.org/10.1146/annurev-neuro-090919-022842
2. Byrne JH, Hawkins RD (2015) Nonassociative learning in invertebrates. Cold Spring Harb Perspect Biol 7(5). https://doi.org/10.1101/cshperspect.a021675
3. Hawkins RD, Byrne JH (2015) Associative learning in invertebrates. Cold Spring Harb Perspect Biol 7(5). https://doi.org/10.1101/cshperspect.a021709
4. Ardiel EL, Rankin CH (2010) An elegant mind: learning and memory in *Caenorhabditis elegans*. Learn Mem 17(4):191–201. https://doi.org/10.1101/lm.960510
5. Rahmani A, Chew YL (2021) Investigating the molecular mechanisms of learning and memory using *Caenorhabditis elegans*. J Neurochem 159(3):417–451. https://doi.org/10.1111/jnc.15510
6. Corsi AK, Wightman B, Chalfie M (2015) A transparent window into biology: a primer on *Caenorhabditis elegans*. Genetics 200(2):387–407. https://doi.org/10.1534/genetics.115.176099
7. Nance J, Frokjaer-Jensen C (2019) The *Caenorhabditis elegans* transgenic toolbox. Genetics 212(4):959–990. https://doi.org/10.1534/genetics.119.301506
8. Stiernagle T (2006) Maintenance of *C. elegans*. In: Community TCeR, editor. WormBook: WormBook
9. Fang-Yen C, Alkema MJ, Samuel AD (2015) Illuminating neural circuits and behaviour in *Caenorhabditis elegans* with optogenetics. Philos Trans R Soc Lond Ser B Biol Sci 370(1677):20140212. https://doi.org/10.1098/rstb.2014.0212
10. Kerr RA (2006) Imaging the activity of neurons and muscles. WormBook 1–13. https://doi.org/10.1895/wormbook.1.113.1
11. Saeki S, Yamamoto M, Iino Y (2001) Plasticity of chemotaxis revealed by paired presentation of a chemoattractant and starvation in the nematode *Caenorhabditis elegans*. J Exp Biol 204 (Pt 10):1757–1764. https://doi.org/10.1242/jeb.204.10.1757
12. Torayama I, Ishihara T, Katsura I (2007) *Caenorhabditis elegans* integrates the signals of butanone and food to enhance chemotaxis to butanone. J Neurosci 27(4):741–750. https://

doi.org/10.1523/JNEUROSCI.4312-06.2007

13. Stein GM, Murphy CT (2014) *C. elegans* positive olfactory associative memory is a molecularly conserved behavioral paradigm. Neurobiol Learn Mem 115:86–94. https://doi.org/10.1016/j.nlm.2014.07.011
14. Fadda M, De Fruyt N, Borghgraef C, Watteyne J, Peymen K, Vandewyer E et al (2020) NPY/NPF-related neuropeptide FLP-34 signals from serotonergic neurons to modulate aversive olfactory learning in *Caenorhabditis elegans*. J Neurosci 40(31): 6018–6034. https://doi.org/10.1523/JNEUROSCI.2674-19.2020
15. Kauffman A, Parsons L, Stein G, Wills A, Kaletsky R, Murphy C. *C. elegans* positive butanone learning, short-term, and long-term associative memory assays. J Vis Exp. 2011(49). https://doi.org/10.3791/2490
16. Bargmann CI (2006) Chemosensation in *C. elegans*. WormBook 1–29. https://doi.org/10.1895/wormbook.1.123.1
17. Sagasti A, Hisamoto N, Hyodo J, Tanaka-Hino M, Matsumoto K, Bargmann CI (2001) The CaMKII UNC-43 activates the MAPKKK NSY-1 to execute a lateral signaling decision required for asymmetric olfactory neuron fates. Cell 105(2):221–232. https://doi.org/10.1016/s0092-8674(01)00313-0
18. Wes PD, Bargmann CI (2001) *C. elegans* odour discrimination requires asymmetric diversity in olfactory neurons. Nature 410(6829):698–701. https://doi.org/10.1038/35070581
19. Bargmann CI, Hartwieg E, Horvitz HR (1993) Odorant-selective genes and neurons mediate olfaction in *C. elegans*. Cell 74(3): 515–527. https://doi.org/10.1016/0092-8674(93)80053-h
20. Tsunozaki M, Chalasani SH, Bargmann CI (2008) A behavioral switch: cGMP and PKC signaling in olfactory neurons reverses odor preference in *C. elegans*. Neuron 59(6): 959–971. https://doi.org/10.1016/j.neuron.2008.07.038
21. Kodama E, Jurado P (2007) Butanone: the memory of a scent. J Neurosci 27(20): 5267–5268. https://doi.org/10.1523/JNEUROSCI.1322-07.2007
22. Nagashima T, Iino Y, Tomioka M (2019) DAF-16/FOXO promotes taste avoidance learning independently of axonal insulin-like signaling. PLoS Genet 15(7):e1008297. https://doi.org/10.1371/journal.pgen.1008297
23. Baugh LR, Hu PJ (2020) Starvation responses throughout the *Caenorhabditis elegans* life cycle. Genetics 216(4):837–878. https://doi.org/10.1534/genetics.120.303565
24. Lim JP, Fehlauer H, Das A, Saro G, Glauser DA, Brunet A et al (2018) Loss of CaMKI function disrupts salt aversive learning in *C. Elegans*. J Neurosci 38(27):6114–6129. https://doi.org/10.1523/JNEUROSCI.1611-17.2018
25. Suzuki H, Thiele TR, Faumont S, Ezcurra M, Lockery SR, Schafer WR (2008) Functional asymmetry in *Caenorhabditis elegans* taste neurons and its computational role in chemotaxis. Nature 454(7200):114–117. https://doi.org/10.1038/nature06927
26. White JG, Southgate E, Thomson JN, Brenner S (1986) The structure of the nervous system of the nematode *Caenorhabditis elegans*. Philos Trans R Soc Lond Ser B Biol Sci 314(1165): 1–340
27. Tomioka M, Adachi T, Suzuki H, Kunitomo H, Schafer WR, Iino Y (2006) The insulin/PI 3-kinase pathway regulates salt chemotaxis learning in *Caenorhabditis elegans*. Neuron 51(5):613–625. https://doi.org/10.1016/j.neuron.2006.07.024
28. Ohno H, Sakai N, Adachi T, Iino Y (2017) Dynamics of presynaptic diacylglycerol in a sensory neuron encode differences between past and current stimulus intensity. Cell Rep 20(10):2294–2303. https://doi.org/10.1016/j.celrep.2017.08.038
29. Satoh Y, Sato H, Kunitomo H, Fei X, Hashimoto K, Iino Y (2014) Regulation of experience-dependent bidirectional chemotaxis by a neural circuit switch in *Caenorhabditis elegans*. J Neurosci 34(47):15631–15637. https://doi.org/10.1523/JNEUROSCI.1757-14.2014
30. Gray JM, Hill JJ, Bargmann CI (2005) A circuit for navigation in *Caenorhabditis elegans*. Proc Natl Acad Sci U S A 102(9):3184–3191. https://doi.org/10.1073/pnas.0409009101
31. Ogg S, Ruvkun G (1998) The C. *elegans* PTEN homolog, DAF-18, acts in the insulin receptor-like metabolic signaling pathway. Mol Cell 2(6):887–893. https://doi.org/10.1016/s1097-2765(00)80303-2
32. Lackner MR, Nurrish SJ, Kaplan JM (1999) Facilitation of synaptic transmission by EGL-30 Gqalpha and EGL-8 PLCbeta: DAG binding to UNC-13 is required to stimulate acetylcholine release. Neuron 24(2):335–346. https://doi.org/10.1016/s0896-6273(00)80848-x
33. Brenner S (1974) The genetics of *Caenorhabditis elegans*. Genetics 77(1):71–94. https://doi.org/10.1002/cbic.200300625

Chapter 3

Mechanical Ablation of Larval Zebrafish Spinal Cord

Samuel Henry Crossman, Mitra Amiri Khabooshan, Sebastian-Alexander Stamatis, Celia Vandestadt, and Jan Kaslin

Abstract

Unlike mammals, adult and larval zebrafish exhibit robust regeneration following traumatic spinal cord injury. This remarkable regenerative capacity, combined with exquisite imaging capabilities and an abundance of powerful genetic techniques, has established the zebrafish as an important vertebrate model for the study of neural regeneration. Here, we describe a protocol for the complete mechanical ablation of the larval zebrafish spinal cord. With practice, this protocol can be used to reproducibly injure upward of 100 samples per hour, facilitating the high-throughput screening of factors involved in spinal cord regeneration and repair.

Key words Spinal cord injury, Tungsten needle fabrication, Zebrafish, Regeneration

1 Introduction

Mammalian spinal trauma is associated with permanent paralysis below the site of injury and lifelong physical disability [1]. This lack of functional regeneration can be attributed to a limited ability to restore damaged cells after injury and the presence of fibrotic scar tissue, which inhibits the regrowth of axons [2–5]. In contrast, some vertebrates, including teleost fish and salamanders, display robust regeneration of cells and axons upon traumatic injury to the spinal cord [6–9]. Adult zebrafish, for example, recover from mechanical spinal cord transections over the course of 4–6 weeks, and studies on regenerating adult zebrafish have uncovered a multitude of factors required for spinal regeneration to occur [10–16]. However, the adult zebrafish spinal column is surrounded by thick layers of muscle and skin, which renders the tissue inaccessible

The original version of the chapter has been revised. A correction to this chapter can be found at https://doi.org/10.1007/978-1-0716-3585-8_19

Sebastian Dworkin (ed.), *Neurobiology: Methods and Protocols*, Methods in Molecular Biology, vol. 2746, https://doi.org/10.1007/978-1-0716-3585-8_3,

for live imaging and requires time-consuming surgery in order to introduce accurate and reproducible lesions. In addition, the relatively slow maturation rate of zebrafish, which require a minimum of 3 months to reach sexual maturity, makes it difficult to conduct studies on adult samples with high throughput.

Like their adult counterparts, larval zebrafish are also capable of recovering from spinal cord injuries, providing an attractive alternative for speculative experiments that require a large number of samples [17, 18]. Importantly, the general function and cellular architecture of the spinal cord are already well developed in the larval zebrafish, allowing direct comparisons to be drawn with adult tissues [19]. The transparency of larval zebrafish also lends itself well to noninvasive live imaging, which can be combined with genetically encoded reporter constructs and CRISPR-/Cas9-mediated F0 knockout approaches or pharmacological drugs to rapidly screen candidate pathways for their involvement in regeneration and repair. Here, we describe an efficient and reproducible protocol for the complete mechanical ablation of larval zebrafish spinal cords. This protocol covers the fabrication of tungsten injury needles and describes the processes required to mount and injure zebrafish larvae with low inter-sample variability and high survival rates. With practice, upward of 100 samples can be processed in an hour, allowing for high-throughput downstream analyses and the rapid discovery of candidate factors involved in spinal regeneration.

2 Materials

Prepare and store all reagents at room temperature unless indicated otherwise. Aqueous solutions should be prepared with ultra-pure distilled water (e.g., Milli-Q). Waste should be disposed of in accordance with institutional policies, and appropriate PPE should be worn at all times.

2.1 Tungsten Dissecting Needles

1. Wire cutters.
2. 5 M aqueous NaOH solution.
3. 50 mL centrifuge tube (e.g., Falcon).
4. Tungsten wire 0.38 mm diameter.
5. Continuous low-voltage power supply (12 V).
6. Crocodile clips ×2.
7. Graphite electrode.
8. 80% ethanol.
9. Reusable pressure-sensitive adhesive (e.g., Blu Tack).
10. 90 mm Petri dish.

2.2 Larval Mounting

1. 1× Ringer's solution: 6.8 g NaCl, 218 mg KCl, and 200 mg $CaCl_2$ in 950 mL of deionized water. Add 5 mL of 1 M HEPES solution and top up to 1000 mL with deionized water. Adjust to pH 7.3–7.4 and autoclave prior to use.
2. Laboratory microwave.
3. Agarose MEEO ultra-quality.
4. 90 mm Petri dishes.
5. Low melting point agarose.
6. 2 mL microcentrifuge tubes.
7. Laboratory heat block or small water bath.
8. 0.168 mg/mL ethyl-m-aminobenzoate methanesulfonate (Tricaine) in Ringer's solution.
9. 1000 μL variable volume pipette.
10. 1000 μL pipette tip.
11. Scalpel blade.
12. Stereomicroscope.
13. Plastic manipulator for gently moving/orienting larva, e.g., an Eppendorf Microloader microcapillary tip with the distal 1–2 cm removed.
14. 10,000 U/mL penicillin/streptomycin.

2.3 Larval Injuries

1. Stereomicroscope.
2. Micro-Tec PV1 precision probe holder for probe diameters 0.1–0.65 mm.
3. Jewelers forceps (Dumont #5).

3 Methods

Conduct all procedures at room temperature unless specified otherwise.

3.1 Preparation of Tungsten Dissecting Needles

1. Use wire cutters to cut 0.38 mm tungsten wire into 4 cm sections.
2. Secure 50 mL centrifuge tube by placing into a suitable tube rack and fill with 5 M NaOH. The tube should be as full as practicably possible without overflowing.
3. Ensure continuous low-voltage power supply is switched off before handling electrodes.
4. Using first crocodile clip, attach graphite electrode to the negative terminal of power supply. Submerge the electrode in the prepared 5 M NaOH solution.

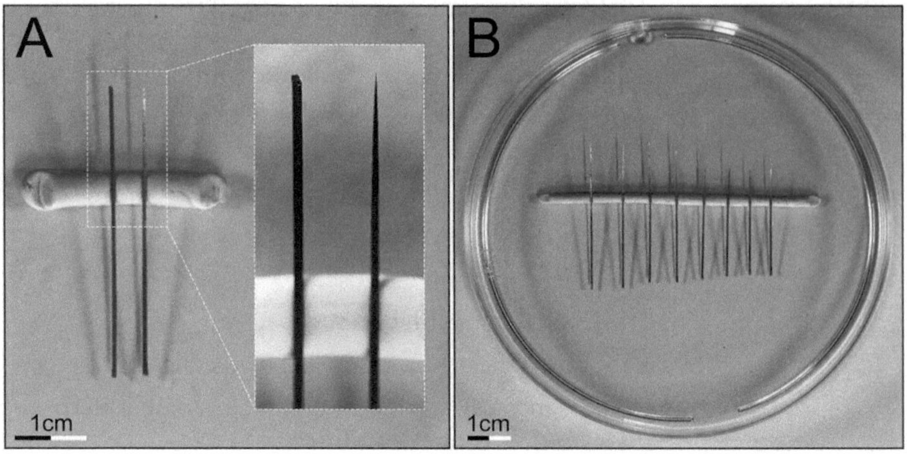

Fig. 1 Fabrication and storage of tungsten injury needles. (**a**) 4 cm tungsten injury sections before (left) and after (right) electrolytic sharpening. Note continuous taper and fine needle tip. (**b**) Lidded Petri dish with thin cylinder of reusable contact adhesive for storage of sharpened injury needles

5. Using second crocodile clip, attach the precut tungsten wire section to the positive terminal of power supply.
6. Switch on power supply.
7. Hold the tungsten section upright and slowly lower into the 5 M NaOH solution, taking care not to touch the two electrodes together. Once partially submerged (approximately 1–2 cm), immediately raise the tungsten section from the 5 M NaOH (*see* **Note 1**). Repeat this motion to create a tapered tip (Fig. 1a, *see* **Note 2**).
8. Once appropriately sharpened, rinse needle thoroughly in 80% ethanol, and store in lidded Petri dish on top of a thin strip of reusable pressure-sensitive adhesive to protect needle tip (Fig. 1b).
9. Repeat **steps 2–8** with a fresh 4 cm tungsten wire section until the desired number of dissecting needles have been made.

3.2 Larval Mounting

1. Pre-warm heat block or small water bath to 42 °C.
2. Prepare mounting plates by mixing 2 g of MEEO ultra-quality agarose with 50 mL Ringer's solution in a 100 mL conical flask to make a 2% agarose solution (*see* **Note 3**). Microwave until a clear and colorless solution is obtained, swirling the flask regularly to ensure even heat distribution. Evenly distribute molten 2% agarose solution between five prepared 90 mm Petri dishes. Care should be taken to minimize bubbles and ensure the entire surface of the dish is covered. Allow plates to cool to room temperature (*see* **Note 4**).

3. Prepare mounting agarose by mixing 100 mg of low melting point (LMP) agarose with 10 mL Ringer's solution in a 100 mL conical flask. Microwave until a clear and colorless solution is obtained, swirling the flask at regular intervals to ensure even heat distribution. Aliquot the molten 1% LMP agarose solution into 10 prepared 2 mL microcentrifuge tubes. Transfer to 42 °C heat block and allow temperature to equilibrate prior to use (*see* **Note 5**).
4. Use a scalpel to remove approximately 1 mm from the distal tip of a 1000 μL pipette tip before securing to a 1000 μL variable volume pipette for transfer of larvae.
5. Anesthetize 3 dpf larvae in Ringer's solution containing approximately 0.168 mg/mL tricaine. Allow 2 min for anesthetic to take effect, there should be no discernible fin movement, and larvae should no longer respond to a gentle tap to the side of the dish.
6. Collect between one and five anesthetized larvae (*see* **Note 6**) with the prepared 1000 μL variable volume pipette and hold vertically to allow larvae to sink to the bottom of the tip with gravity. Place the tip into the molten LMP agarose and transfer larvae, taking care to avoid any dilution of the agarose with media. Remove tip and expel any remaining media.
7. Gently pipette LMP agarose solution up and down in order to mix the larvae with the mounting substrate before collecting the larvae again in the tip of the pipette. Expel collected larvae onto the mounting plate within a small droplet of LMP agarose (*see* **Note 7**). The droplet should be positioned roughly halfway between the center and the edge of the dish to ensure that larvae are accessible during subsequent injury stages (Fig. 2a). LMP agarose should be returned to 42 °C heat block when not in use to maintain molten state.
8. Rapidly transfer dish to stereomicroscope and adjust focus and magnification to visualize droplet. Use a plastic manipulator to quickly orientate each larvae sagittally within the droplet so that the head is facing left (*see* **Note 8**) and the dorsal surface is facing the experimenter (Fig. 2a, *see* **Note 9**).
9. Rotate the mounting dish approximately 45° clockwise before repeating **steps 6–8**. Repeat until the entire circumference of the dish is occupied (Fig. 2b).
10. Allow 1–2 min for the final droplet of agarose to set before covering the samples with a thin layer of Ringer's solution containing 0.168 mg/mL tricaine and 100 U/mL penicillin/streptomycin.

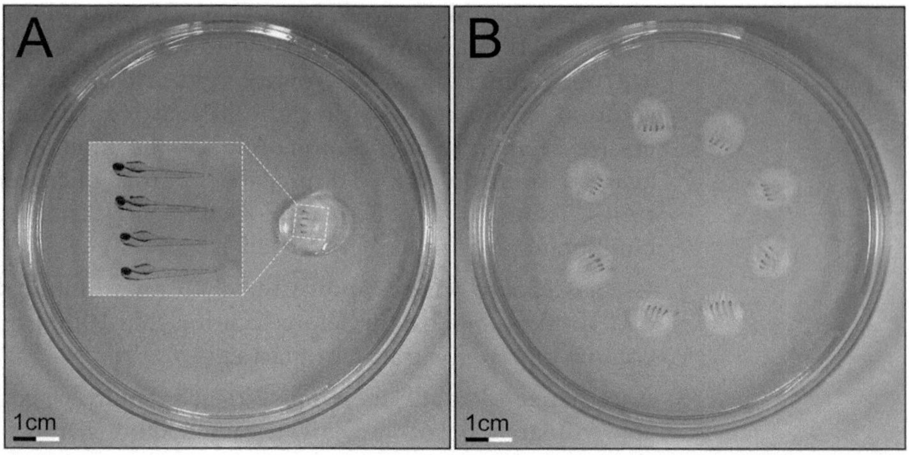

Fig. 2 Mounting and orientation of larval samples. (**a**) 1–5 samples are transferred to mounting plates in a droplet of molten low melt agarose. Samples are positioned roughly midway between the center and the right-most edge of the dish, with the dorsal surface facing the experimenter and the head facing left toward the center of the dish (**inset**). Left-handed individuals should reverse this orientation, positioning the droplet on the left-hand side of the plate with the dorsal surface facing the experimenter and the head facing to the right. (**b**) A fully mounted injury plate, with 32 larvae mounted in 8 droplets

3.3 Larval Injury

1. Securely fasten tungsten dissecting needle (prepared as described in Subheading 3.1) into probe holder (*see* **Note 10**).
2. Using a stereomicroscope to visualize samples, rotate the mounting dish so that the larvae within a single droplet are visible and oriented perpendicular to the direction of the needle with the dorsal surface facing the experimenter.
3. Use needle to gently scratch the dorsal surface of the larvae roughly opposite the urogenital pore, breaking the skin above the intended injury sight (*see* **Note 11**).
4. Insert tip into the spinal cord of the larvae, aiming for an insertion site approximately halfway between the notochord and the dorsal epidermis (Fig. 3a, b). Retract the needle in a ventral to dorsal direction, scooping out tissue in a single continuous motion. Repeat approximately three times, inserting the needle deeper into the dorsal tissue on each occasion, until the needle has reached the dorsal surface of the notochord. Take care to remove as much tissue as possible without inflicting any damage to the notochord itself (Fig. 3c). Injury size should correspond to the width of approximately one larval somite (100–150 μm).
5. Repeat **steps 3–4** for each larva within the droplet before rotating the dish and repeating the process for each droplet on the plate.

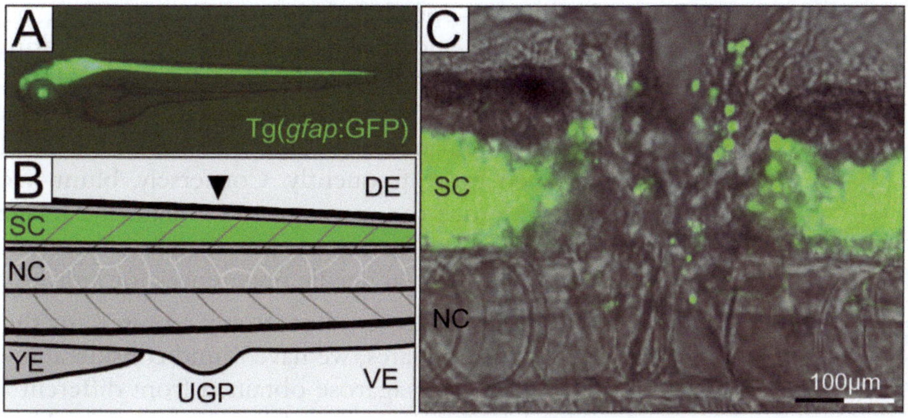

Fig. 3 Larval injury. (**a**) The larval spinal cord, as visualized with the pan-glial transgene Tg(*gfap*:GFP). (**b**) Schematic representation of the larval spinal cord and its associated structures. The optimal injury site, caudal to the yolk extension and roughly opposite the urogenital pore, is denoted with black arrowhead. SC = spinal cord, NC = notochord, YE = yolk extension, UGP = urogenital pore, DE = dorsal epidermis, and VE = ventral epidermis. (**c**) Example spinal cord injury. All spinal cord tissue has been removed across a 100–150 μm region, but the underlying notochord remains undamaged and intact. Green = Tg(*gfap*:GFP). Reproduced from Vandesdadt et al., 2021 with permission from Elsevier

6. Use jewelers forceps to demount injured larvae from the plate. First, remove agarose surrounding the head of the larvae before gently running the forceps over the top of the sample to create a groove. Next, insert the forceps below the larvae and gently push in an upward direction to release the sample from the agarose. Repeat demounting process for each sample on the dish.
7. Use a 1000 μL variable volume pipette to transfer injured larvae to a fresh 90 mm Petri dish containing Ringer's solution (without tricaine) and 100 U/mL penicillin/streptomycin, taking care to minimize transfer of media from the injury plate. Larvae that have incurred noticeable damage to the notochord should be discarded due to reduced viability and delayed regeneration. Repeat for all larvae on the plate.
8. Remove needle from probe holder, sterilize with 80% ethanol, wash in deionized water, and store for further use (*see* **Note 12**).

4 Notes

1. During needle making, the tungsten section should never be held stationary whilst the current is running and the needle is submerged in NaOH. Continuous motion of the submerged wire ensures a smooth taper to the needle, which is essential for needle strength and longevity.

2. By altering the speed and depth of submersion, needles of varying lengths and sharpness can be created. During injury, sharper needles are easier to break the skin with but are more difficult to wield due to their increased flexibility. Fine-tipped needles also damage easily and will need to be replaced or re-sharpened more frequently. Conversely, blunt needles are more durable but can cause greater collateral damage to surrounding tissue. Independent optimization is recommended to determine the optimal needle preferences for individual users.
3. While ultrapure agarose is not strictly required for the fabrication of mounting plates, we have found considerable variation in the properties of agarose obtained from different suppliers when used for the purpose of larval mounting. Many of the lower grades of agarose that we have tested display poor adhesion between the mounting plate and the LMP agarose mounting media, resulting in detachment of larvae and difficulties during subsequent orientation and injury steps. We have found that the Carl Roth MEEO ultra-quality agarose described in this protocol displays the best mechanical properties of the various products that we have tested in the lab.
4. Unused mounting plates can be wrapped tightly in plastic wrap and stored at 4 °C for up to a week.
5. Unused aliquots of mounting agarose can be stored in an airtight container at 4 °C for up to 6 weeks and can be melted again by placing in a heat block set to 72 °C for 5–10 min prior to use.
6. With practice, more larvae can be mounted within each droplet. It is recommended that beginners start with a single larva and focus on correct orientation before increasing larval numbers. With experience, five or more larvae can be mounted within a single droplet.
7. It is important that droplets of agarose are of sufficient size to ensure that the larvae are firmly secured in place. Smaller droplets of agarose will also set more quickly, which reduces the amount of time available for larval orientation.
8. Left-handed individuals should reverse this orientation and mount samples so that the head is facing right. In both instances, the dorsal surface needs to face the experimenter.
9. Successful orientation is key to a successful injury. If larvae are not mounted perpendicularly to the mounting plate, it is very hard to conduct clean injuries without damaging the notochord or neighboring segments. It is always better to mount fewer larvae in the correct orientation than to have more samples on a plate that are poorly aligned and thus difficult to injure precisely.

10. Length of needle protruding from the probe holder can have drastic effects on the ease of injury. Too long and the needle will be too flexible; too short and access to the sample is occluded. An exposed needle length of approximately 2 cm is optimal.
11. Breaking the skin before inserting the needle into the ventral spinal cord reduces the risk of tissue tearing during injury and helps prevent collateral damage to neighboring segments.
12. Needles can be reused for multiple rounds of injuries if stored correctly. Used needles typically need to be resharpened prior to their subsequent use by following **steps 2–8** of Subheading 3.1 of this protocol (Preparation of Tungsten Dissecting Needles). It is typically quicker to resharpen needles than it is to make them from scratch as the broad profile of the needle has already been established. As a result, 2–3 rounds of electrolyte submersion will usually be sufficient. Resharpening reduces the length of a needle, and needles can be resharpened until they are too short to be mounted into the probe holder. Any needles below 2 cm should be discarded in an appropriate sharp waste container.

References

1. Alizadeh A, Dyck SM, Karimi-Abdolrezaee (2019) Traumatic spinal cord injury: an overview of pathophysiology, models and acute injury mechanisms. Front Neurol 10:282
2. Houweling DA, Bär PR, Gispen WH, Joosten EA (1998) Spinal cord injury: bridging the lesion and the role of neurotrophic factors in repair. Prog Brain Res 117:455–471
3. Hara M, Kobayakawa K, Ohkawa Y, Kumamaru H, Yokota K, Saito T, Kijima K, Yoshizaki S, Harimaya K, Nakashima Y, Okada S (2017) Interaction of reactive astrocytes with type I collagen induces astrocytic scar formation through the integrin-N-cadherin pathway after spinal cord injury. Nat Med 23(7):818–828
4. Dias DO, Kim H, Holl D, Werne Solnestam B, Lundeberg J, Carlén M, Göritz C, Frisén J (2018) Reducing pericyte-derived scarring promotes recovery after spinal cord injury. Cell 173(1):153–165
5. Bradbury EJ, Burnside ER (2019) Moving beyond the glial scar for spinal cord repair. Nat Commun 10(1):3879
6. Bernstein JJ (1964) Relation of spinal cord regeneration to age in adult goldfish. Exp Neurol 9:161–174
7. Butler EG, Ward MB (1967) Reconstitution of the spinal cord after ablation in adult Triturus. Dev Biol 15(5):464–486
8. Tanaka EM, Ferretti P (2009) Considering the evolution of regeneration in the central nervous system. Nat Rev Neurosci 10(10): 713–723
9. Becker CG, Becker T (2015) Neuronal regeneration from ependymo-radial glial cells: cook, little pot, cook! Dev Cell 32(4):516–527
10. Becker T, Wullimann MF, Becker CG, Bernhardt RR, Schachner M (1997) Axonal regrowth after spinal cord transection in adult zebrafish. J Comp Neurol 377(4):577–595
11. van Raamsdonk W, Maslam S, de Jong DH, Smit-Onel MJ, Velzing E (1998) Long term effects of spinal cord transection in zebrafish: swimming performances, and metabolic properties of the neuromuscular system. Acta Histochem 100(2):117–131
12. Reimer MM, Kuscha V, Wyatt C, Sörensen I, Frank RE, Knüwer M, Becker T, Becker CG (2009) Sonic hedgehog is a polarized signal for motor neuron regeneration in adult zebrafish. J Neurosci 29(48):15073–15082
13. Goldshmit Y, Sztal TE, Jusuf PR, Hall TE, Nguyen-Chi M, Currie PD (2012)

Fgf-dependent glial cell bridges facilitate spinal cord regeneration in zebrafish. J Neurosci 32(22):7477–7492

14. Dias TB, Yang YJ, Ogai K, Becker T, Becker CG (2012) Notch signaling controls generation of motor neurons in the lesioned spinal cord of adult zebrafish. J Neurosci 32(9): 3245–3252
15. Mokalled MH, Patra C, Dickson AL, Endo T, Stainier DY, Poss KD (2016) Injury-induced ctgfa directs glial bridging and spinal cord regeneration in zebrafish. Science 354(6312): 630–634
16. Goldshmit Y, Tang JKKY, Siegel AL, Nguyen PD, Kaslin J, Currie PD, Jusuf PR (2018) Different Fgfs have distinct roles in regulating neurogenesis after spinal cord injury in zebrafish. Neural Dev 13(1):24
17. Briona LK, Poulain FE, Mosimann C, Dorsky RI (2015) Wnt/ß-catenin signaling is required for radial glial neurogenesis following spinal cord injury. Dev Biol 403(1):15–21
18. Vandestadt C, Vanwalleghem GC, Khabooshan MA, Douek AM, Castillo HA, Li M, Schulze K, Don E, Stamatis SA, Ratnadiwakara M, Änkö ML, Scott EK, Kaslin J (2021) RNA-induced inflammation and migration of precursor neurons initiates neuronal circuit regeneration in zebrafish. Dev Cell 56(16):2364–2380
19. Alper SR, Dorsky RI (2022) Unique advantages of zebrafish larvae as a model for spinal cord regeneration. Front Mol Neurosci 15: 983336

Chapter 4

High-Throughput Multiplex Immunohistochemistry of Glioma Organoids

Marija Dinevska, Samuel S. Widodo, and Theo Mantamadiotis

Abstract

The invasive capacity and progression of glioblastoma cells and neoplastic cells in other are dependent on interactions with the surrounding tumor microenvironment. In particular, cancer cells form a reciprocal relationship with noncellular dysregulated extracellular matrix in the tumors. Here, we describe a protocol that can be used to model the functional relationship between tumor cells and extracellular matrix. We demonstrate how 3D organoids, including glioma tumor organoids, can be processed, embedded, and sectioned in a high-throughput setup that enables investigation of the organoids by histopathological methods, multiplex immunohistochemistry, and spatial analysis within the same section.

Key words Glioma, Glioblastoma, Organoid, Multiplex immunohistochemistry, Spheroid microarray

1 Introduction

Glioblastoma (GBM), a grade 4 tumor, is the most common and malignant cancer affecting the adult central nervous system [1]. Despite advances in targeted therapy and immunotherapy, GBM remains the most lethal neurological malignancy, with a median survival of 16 months [2]. A key feature of GBM that contributes to poor patient prognosis is the highly infiltrative nature of GBM cells and their ability to invade and colonize the surrounding healthy brain tissue [3]. For GBM cells to acquire motility, they need to secrete proteases capable of degrading the surrounding extracellular matrix (ECM) to form a path for the invading cells [4, 5]. During cancer development, the ECM is disrupted, which contributes to the establishment of a pro-tumorigenic tumor microenvironment (TME) [6]. The ECM is integral in modulating cellular interactions and cell signaling events that regulate functions, including cell proliferation, cell differentiation, and cell invasion [7]. Understanding how tumor cells interact with the TME is critical to identifying cancer cell

Sebastian Dworkin (ed.), *Neurobiology: Methods and Protocols*, Methods in Molecular Biology, vol. 2746, https://doi.org/10.1007/978-1-0716-3585-8_4,

oncogenic mechanisms, as these interactions and functional communications are not unidirectional but rather a complex interdependent relationship between the cells and the ECM [6]. With this in mind, 3D organoid and spheroid models are biologically relevant for interrogating how tumor cells behave and functionally interact with their surrounding microenvironment, in particular, how tumor cells interact with the surrounding ECM to facilitate invasion, as well as how ECM composition influences invasion.

The protocol presented herein describes a high-throughput method to investigate the interaction between 3D organoids/spheroids and the ECM, with the capacity to screen for drugs and other treatments that regulate cell invasion. The protocol described is amenable to modification to allow investigation of individual or specific combinations of ECM proteins and other ECM properties, including biophysical properties such as stiffness. The protocol allows for downstream analysis of spheroids using histological analyses, including molecular profiling by single or multiplexed immunohistochemistry (mIHC) and spatial analysis, since the architecture of the spheroids and ECM-interacting cells is preserved. mIHC is advantageous over single antibody immunohistochemistry because comprehensive analysis of the co-expression and spatial organization of multiple biomarkers within a single spheroid section can be performed [8]. In the protocol presented here, we use GBM spheroids to identify the signaling status of cells to illustrate how this methodological pipeline can be used to examine cellular activation. However, the protocol can be used for any cell type, which can be cultured in 3D, and analysis is only limited by the availability of antibodies to identify biomarkers of interest.

2 Materials

2.1 Cell Culture

1. Spheroid growth medium: Dulbecco's Modified Eagle Medium (DMEM)/Nutrient Mixture F-12, 1 × B27 supplement (preferably vitamin-A free), basic fibroblast growth factor (FGF) 10 ng/mL, epidermal growth factor (EGF) 20 ng/mL. Store at 4 °C.
2. Adherent cell medium: DMEM supplemented with 1× GlutaMAX, 1× antibiotic-antimycotic, and 10% fetal bovine serum (FBS). Store at 4 °C.
3. Accutase cell detachment solution.
4. 1× Phosphate buffered saline (PBS). pH 7.4: 13.65 g/L Na_2HPO_4 (anhydrous), 2.4 g/L KH_2PO_4 (anhydrous), 80 g/L KCl, NaOH to pH 7.4
5. 6-Well ultralow attachment tissue culture plate

6. Tissue culture-treated vented flasks.
7. Trypsin-ethylenediaminetetraacetic acid (EDTA).

2.2 3D Invasion Assay

1. Spheroid growth medium: DMEM/F—12 medium, 1 × B27 supplement, basic FGF 10 ng/mL, EGF 20 ng/mL. Store at 4 °C and use within 1 month.
2. Cultrex spheroid invasion matrix. Store at −80 °C. Avoid freeze-thawing.
3. 96-Well round bottom ultralow attachment plate.

2.3 Spheroid Microarray

1. 10% Neutral buffered formalin (NBF)
2. Richard-Allan Scientific HistoGel™ (ThermoFisher Scientific). Store at 4 °C.
3. Cryomold mold (2.5 × 2.0 × 0.5 cm).
4. Tissue cassette mold (3.0 × 2.5 × 0.4 cm).
5. Superfrost™ Plus glass slides.
6. Paraffin wax.
7. Microtome.

2.4 Multiplex Immunohistochemistry

1. Xylene.
2. Ethanol: 2 × 100%, 70%, and 30% in distilled water.
3. Hydrophobic barrier peroxidase anti-peroxidase (PAP) Pen.
4. Antigen retrieval buffer: citrate buffer (pH 6: pH 6: 24.269 g/L $Na_3C_6H_5O_7$, 3.358 g/L $C_6H_8O_7$, HCL to pH 6) or Tris-EDTA buffer (pH 9: 1.21 g/L $C_4H_{11}NO_3$, 0.337 g/L $C_{10}H_{16}N_2O_8$).
5. 3% Hydrogen peroxide: Add 1 mL of 30% hydrogen peroxide to 9 mL of distilled water
6. Phosphate-Buffered Saline + Tween-20 (PBST): 50 μL Tween-20 in 50 mL PBS.
7. Opal 6-plex Manual Detection Kit (Akoya Biosciences): 1 × antibody diluent, 1 × Opal anti-mouse + Rb HRP secondary, opal-reactive fluorophores (520, 540, 570, 620, 650, 690) and 10 × spectral 4′,6-diamidino-2-phenylindole (DAPI) (diluted to 1 × in PBST).
8. Glass coverslips (24 × 50 mm).
9. ProLong Gold Antifade Mountant without DAPI. Store at 4 °C.

2.5 Equipment

1. Cell incubator with CO_2 supply (any type suitable for culture of mammalian cells).
2. Centrifuge with swinging-bucket rotor and microplate holder.

3. Vectra 3 or Vectra Polaris widefield multispectral microscope (or suitable equivalent for imaging of Opal fluorophores).
4. Inverted brightfield microscope with camera.
5. Fiji ImageJ.

3 Methods

3.1 Cell Culture

For non-adherent cells:

1. Seed and grow cells in ultralow attachment 6-well plates in 3 mL of spheroid growth medium at 37 °C and 10% atmospheric CO_2 in a humidified incubator until spheroids reach the desired size (~3–4 days; *see* **Note 1**).
2. Collect spheroids in a 15 mL Falcon tube and centrifuge at 120× *g* for 5 min, at ambient/room temperature.
3. Remove the supernatant, add 500 μL of accutase for 5 min, and place into a 37 °C water bath.
4. Triturate several times using a P1000 pipette and add 5 mL of 1 × PBS (*see* **Note 2**).
5. Centrifuge for 5 min at 120× *g*, remove the supernatant, and resuspend in required volume of spheroid growth medium.
6. Seed 3×10^3 cells per well of a 96-well round bottom ultralow attachment plate in a 50 μL cell suspension (*see* **Note 3**).
7. Centrifuge the plate at 200× g for 3 min at room temperature in a swinging-bucket rotor and incubate for 72 h at 37 °C and 10% atmospheric CO_2 in a humidified incubator to promote spheroid formation (*see* **Note 4**). Skip to Subheading 3.2 if using non-adherent cells.

For adherent cells:

1. Grow cells in adherent tissue–culture treated vented flasks in adherent cell medium (see materials) at 37 °C and 10% atmospheric CO_2 in a humidified incubator until 70–80% confluency is reached.
2. Harvest cells by aspirating culture media and gently washing cells with 1 × PBS.
3. Enzymatically detach adherent cells by trypsinization (1–2 mL per T-75 flask) for 5 min at 37 °C.
4. Neutralize trypsin with adherent cell medium and centrifuge cells in a 15 mL Falcon tube at 120× *g* for 5 min.

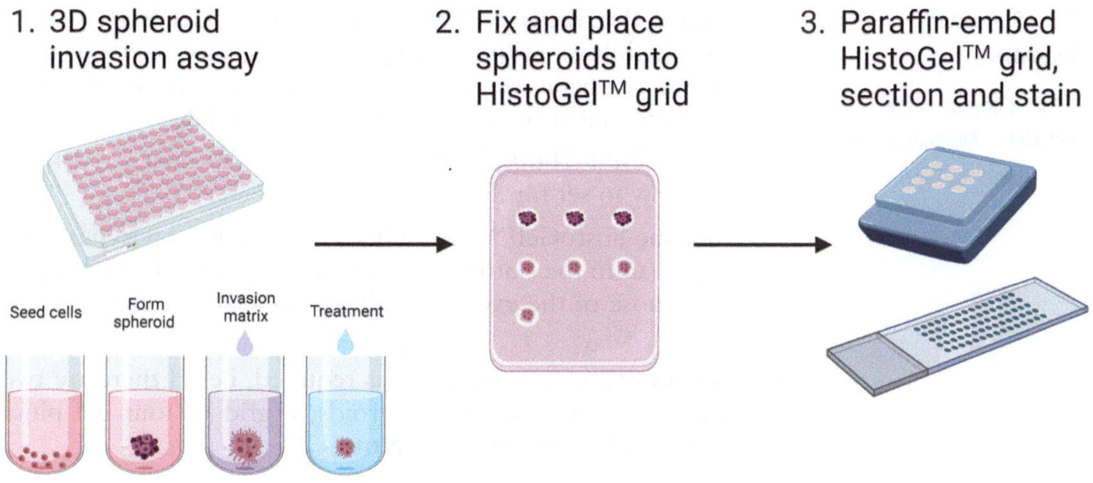

Fig. 1 Development of spheroid microarray grids. Single cells are seeded in a 96-well plate to form spheroids, before the addition of Cultrex invasion matrix. Spheroids are incubated to allow cell invasion over 96 h, with or without treatment, followed by fixation in 10% NBF. Spheroids are individually transferred into a premade HistoGel™ microarray/grid, which is then processed, and paraffin is embedded. Downstream staining with histological stains and multiplex immunohistochemistry is performed. (The figure is created with BioRender)

5. Remove the supernatant and resuspend cells between 5 and 10 mL of 1× PBS before centrifugation at 120× *g* for 5 min (*see* **Note 5**).
6. Resuspend the cells in spheroid growth medium and plate 3×10^3 cells per well in a 96-well round bottom ultralow attachment plate, in a 50 μL cell suspension (*see* **Note 6**).
7. Centrifuge the plate at 200× *g* for 3 min at room temperature in a swinging-bucket rotor and incubate for 72 h at 37 °C and 10% atmospheric CO_2 in a humidified incubator to promote spheroid formation (Fig. 1).

3.2 3D Invasion Assay

1. Following spheroid formation (Subheading 3.1), place 96-well plate on ice in a refrigerator to cool wells for 15 min (*see* **Note 7** and **8**).
2. Add 50 μL of invasion matrix (see materials) per well and centrifuge in a swinging-bucket rotor for 300× *g* at 4 °C for 5 min (*see* **Note 9**).
3. Incubate the plate at 37 °C and 10% atmospheric CO_2 in a humidified incubator for 1 h.
4. Add 100 μL spheroid growth medium containing various inhibitors or chemoattractants (*see* **Note 10**).
5. Incubate the plate at 37 °C and 10% atmospheric CO_2 in a humidified incubator for 72 to 96 h and image every 24 h using the 4× objective (Fig. 1).
6. Image analysis can be performed using Fiji ImageJ to quantify the area/distance of cell invasion (*see* **Note 11**).

3.3 Spheroid Microarray

3.3.1 Generating the HistoGel™ Microarray/Grid

1. Prepare the Richard-Allan Scientific HistoGel™ microarray mold as follows prior to fixing spheroids. HistoGel™ is solid at room temperature and needs to be liquified by microwaving on low for short bursts of 5–10 s (*see* **Note 12**).
2. Transfer 500–1000 μL of liquid HistoGel™ into a cryomold and allow to set for 30 min on ice.
3. Once the HistoGel™ is solidified, use a P5000 tip to puncture the gel (until the bottom of the cryomold is reached) and pick up/dispose of the excess gel plug with a pair of fine tweezers (Fig. 1).
4. This process can be repeated as required, i.e., if there are two control and two treated spheroids, a grid of four gel plugs would need to be made (*see* **Note 13**).

3.3.2 Fixing and Embedding the Spheroids

1. Following 72–96 h of incubation, to allow cells to invade, remove as much of the medium as possible (~150 μL) from each well, without disturbing the spheroid.
2. Add 100 μL of 10% NBF in each well and leave at room temperature for 30 min.
3. Dilute the NBF by adding 100 μL of 1 × PBST to each well.
4. Remove and dispose of ~200 μL of solution from each well. Take care not to remove cells/spheroid.
5. Using a P200 and a fine tweezer or needle, transfer a single spheroid from each well to the HistoGel™ microarray plugs.
6. Repeat this process for the remaining spheroids to set up the HistoGel™ microarray mold.
7. Using 50–100 μL of liquified HistoGel™, fill the gel plugs to form one uniform mold (*see* **Note 14**).
8. Allow the HistoGel™ microarray to solidify at 4 °C overnight.
9. Remove the HistoGel™ microarray from the cryomold by pushing the sides down and place it into the cassette for processing and embedding (*see* **Note 15**).
10. Cut the paraffin block into 4 μm thick sections and place the sections onto Superfrost plus glass slides.

3.4 Multiplex Immunohistochemistry

1. Bake the slides at 60 °C to enhance binding of tissue onto the slides.
2. Deparaffinize the sections by immersing the slides with xylene for 5 min each. Repeat this step once.
3. Rehydrate the slides through a series of decreasing concentrations of ethanol (100%, 100%, 70%, and 30%) for 2 min each.
4. Wash in tap water prior to heat-induced antigen retrieval in antigen retrieval buffer (*see* **Note 16**).

5. Use a hydrophobic barrier PAP Pen to circle around all spheroids, to avoid leakage and/or drying of samples.
6. Block for nonspecific binding of antibodies in 3% H_2O_2 for 10 min and wash three times in 1 × PBST.
7. Block in 2–3 drops of 1 × antibody diluent for 5 min at room temperature.
8. Add 150 μL of diluted primary antibody (dilute in 1 × antibody diluent) per slide for 1 h at room temperature.
9. Tap off the primary antibody and wash three times for 5 min in 1 × PBST.
10. Add 2 or 3 drops of Opal anti-Ms + Rb HRP secondary for 30 min at room temperature. Tap off and wash three times for 5 min in 1 × PBST.
11. Add 150 μL Opal-reactive fluorophore (1:150 diluted in 1 × amplification diluent) for 10 min at room temperature (*see* **Note 17**). Tap off the used solution and wash three times for 5 min in 1 × PBST.
12. For multiplexing, repeat antigen retrieval and steps 7–10 for each primary antibody (*see* **Note 18**).
13. Add 150 μL spectral DAPI (diluted to 1 × in PBST) for 2 min at room temperature. Wash slides three times in PBST.
14. Mount slides with 100 μL ProLong Gold Antifade Mountant and coverslip.
15. Scan the stained slides using Vectra 3 or Vectra Polaris microscopes (Fig. 2).
16. For alternatives to multiplex immunohistochemistry, *see* **Note 19**.

4 Notes

1. This protocol has been optimized for mouse spheroids and human glioblastoma cell lines – LN229 (ATCC CRL-2611), U87MG (ATCC HTB-14), U118 (ATCC HTB-15), as well as primary patient-derived cell lines.
2. Triturate 20–30 times to ensure that spheroids are broken down into single cells.
3. Ensure that cells are seeded in a 96-well round bottom, ultra-low attachment plate, and not a flat bottom plate, otherwise spheroids will not form at the center of the well.
4. Following centrifugation, check that all single cells are in the center of the well to prevent multiple spheroids forming. If cells are not collected in the center of the wells, centrifuge in a swinging-bucket rotor for an additional 3 min at 200× g.

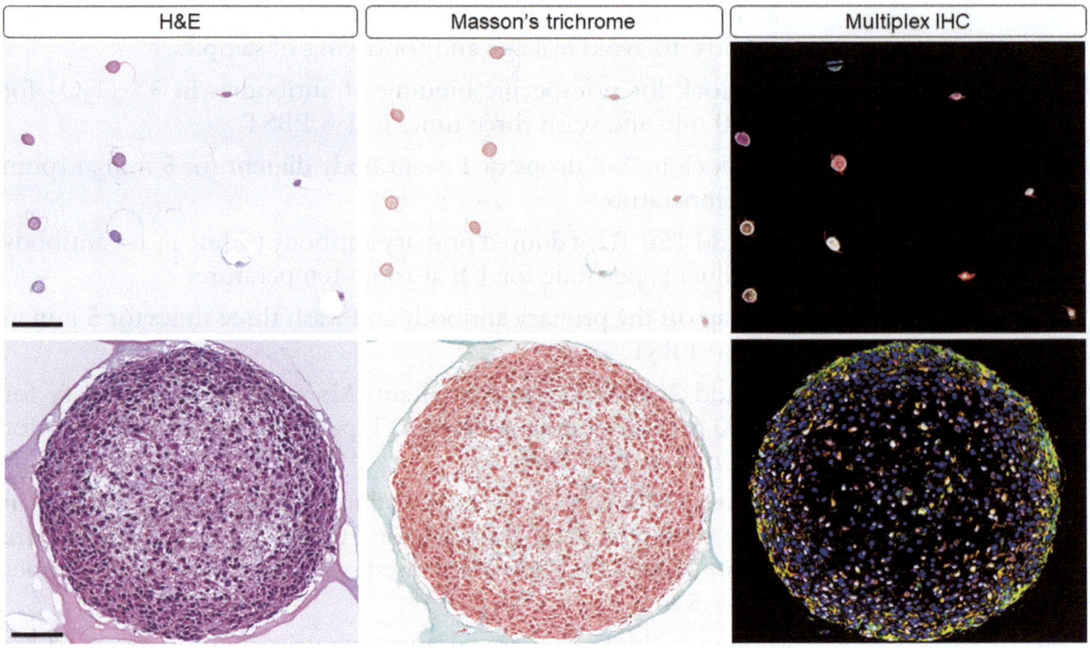

Fig. 2 Staining of spheroid microarrays. Spheroid microarray sections were either stained with hematoxylin and eosin (H&E) and Masson's trichrome or subjected to mIHC using a panel of antibodies (pAKT, pERK, and pCREB). The scale bar for the upper image panel is 1 mm and 100 μm for the lower image panel

5. It is critical to wash cells in 1 × PBS ensuring that the FBS is removed.
6. Optimal seeding density for each cell line should be predetermined.
7. The 3D invasion matrix is not required if it is not part of the research question. The invasion matrix can be replaced with spheroid growth medium.
8. Ensure that the plate is placed on ice for 15 min to prevent rapid matrix polymerization prior to centrifugation.
9. Cultrex invasion matrix can be replaced with other suitable hydrogels, collagen, or other ECM proteins.
10. For this protocol, inhibitors of MAPK and PI3K signaling were used. Other drug treatments or inhibitors, chemoattractants, growth factors, and conditioned media could also be used to assess their impact on cell invasion and migration.
11. Invasion analysis is performed on Fiji ImageJ by using the freehand tool to circle around the invasive edge of the spheroid and measuring the area (analyses > measure). The invasive area following 24-, 48-, 72-, and 96-h invasion is compared to 0 h.
12. Take care when heating the HistoGel™, to prevent overheating and spill over.

13. Note the arrangement of the spheroids in the grid to enable correct identification of each sample, post-staining.
14. Gently and slowly dispense 50–100 μL of liquified HistoGel™ ensuring not to disturb the positioning of the spheroid.
15. If the HistoGel™ gel block is difficult to remove from the cryomold, a scalpel can be used to tease out the gel by running the blade along the cryomold edges.
16. The antigen retrieval buffer to be used can be determined based on data provided from the antibody manufacturer and/or prior optimization.
17. Optimal antibody–fluorophore pairing can be predetermined during singleplex Immunohistochemistry (IHC) and Immunofluorescence (IF) staining.
18. Serial antibody staining can be repeated up to five times for all available fluorophores.
19. Multiplex immunohistochemistry is performed using the Opal 6-plex Manual Detection Kit (Akoya Biosciences) to detect six antibodies of interest on a single slide. If this is not possible, alternatives include chromogenic immunohistochemistry, as well as histological stains including hematoxylin and eosin, or Masson's trichrome.

References

1. Berger, T. R., Wen, P. Y., Lang-Orsini, M., & Chukwueke, U. N. (2022). World Health Organization 2021 classification of central nervous system tumors and implications for therapy for adult-type gliomas: a review. JAMA Oncol, 8(10), 1493–1501.
2. van Solinge TS, Nieland L, Chiocca EA, Broekman MLD (2022) Advances in local therapy for glioblastoma – taking the fight to the tumour. Nat Rev Neurol 18(4):221–236
3. de Gooijer MC, Guillén Navarro M, Bernards R, Wurdinger T, van Tellingen O (2018) An experimenter's guide to glioblastoma invasion pathways. Trends Mol Med 24(9):763–780
4. Hatoum A, Mohammed R, Zakieh O (2019) The unique invasiveness of glioblastoma and possible drug targets on extracellular matrix. Cancer Manag Res 11:1843–1855
5. Paw I, Carpenter RC, Watabe K, Debinski W, Lo H-W (2015) Mechanisms regulating glioma invasion. Cancer Lett 362(1):1–7
6. Cox, T R (2021) The matrix in cancer Nat Rev Cancer, 21(4), 217–238.
7. Bonnans C, Chou J, Werb Z (2014) Remodelling the extracellular matrix in development and disease. Nat Rev Mol Cell Biol 15(12):786–801
8. Widodo SS, Hutchinson RA, Fang Y, Mangiola S, Neeson PJ, Darcy PK et al (2021) Toward precision immunotherapy using multiplex immunohistochemistry and in silico methods to define the tumor immune microenvironment. Cancer Immunol Immunother 70(7):1811–1820

Chapter 5

The Bilateral Carotid Artery Stenosis (BCAS) Model of Vascular Dementia

Quynh Nhu Dinh and Thiruma Arumugam

Abstract

Vascular dementia is the second most common form of dementia after Alzheimer's disease. Chronic cerebral hypoperfusion is a key contributor to the development of vascular dementia. In this chapter, we describe the surgical procedures used for bilateral carotid artery stenosis (BCAS) surgery to induce chronic cerebral hypoperfusion. Mice that undergo BCAS surgery develop the hallmarks of vascular dementia including white matter lesions, neuroinflammation, and cognitive impairment. This technique may be used for studies of chronic cerebral hypoperfusion and vascular dementia in mice.

Key words Vascular dementia, Carotid arteries, Brain, Chronic cerebral hypoperfusion, Cerebral blood flow

1 Introduction

Dementia is an umbrella term used to describe neurological conditions characterized by a severe decline in cognitive function and loss of ability to perform daily activities. It is a major healthcare and economic burden affecting 50 million people worldwide [1]. Vascular dementia is a frequent form of dementia (up to 30% of cases) and the second leading cause after Alzheimer's disease [2]. Vascular dementia is a progressive neurodegenerative disorder associated with chronic cerebral hypoperfusion caused by cerebrovascular disease. Vascular dementia has a complex pathology where cardiovascular comorbidities, including hypertension and stroke, promote the accumulation of vascular injury in the brain that ultimately results in chronic cerebral hypoperfusion [3]. This chronic cerebral hypoperfusion leads to brain energy imbalance, excitotoxicity and disruption of calcium homeostasis, and eventually sustained production of reactive oxygen species, proinflammatory cytokines, and matrix metalloproteinases. This promotes the development of white matter lesions and, ultimately, cognitive

Sebastian Dworkin (ed.), *Neurobiology: Methods and Protocols*, Methods in Molecular Biology, vol. 2746,
https://doi.org/10.1007/978-1-0716-3585-8_5,

impairment [3]. Patients with vascular dementia typically have a shorter life expectancy and poorer prognosis than patients with Alzheimer's disease [4]. Currently, there are no specific or disease-modifying treatments for vascular dementia mainly due to an incomplete understanding of the underlying pathophysiology.

Animal models are useful tools to better understand mechanisms of disease. The bilateral carotid artery stenosis (BCAS) model is widely regarded as the most clinically relevant animal model of vascular dementia [5, 6]. This model involves permanent placement of coils around each carotid artery, restricting blood flow to ~40% of normal levels. This cerebral hypoperfusion is maintained for up to 12 weeks, over which time mice develop hallmarks of vascular dementia including neuroinflammation, white matter lesions, and marked cognitive impairment [7].

In this chapter, we outline the methodology to induce chronic cerebral hypoperfusion using the BCAS model.

2 Materials

1. Microcoils with an internal diameter of 0.18 mm (Sawane Spring Co., Ltd, Japan).
2. Microcoils with an internal diameter of 0.18 mm (Sawane Spring Co., Ltd, Japan).
3. Dissecting microscope.
4. Scissors—straight tip (e.g., Fine Science Tools scissors 14088-10) (*see* **Note 1**).
5. Forceps × 1—straight tip (e.g., Fine Science Tools fine forceps 11254-20).
6. Forceps × 2—curved tip (e.g., Fine Science Tools fine forceps 11274-20).
7. Retractor.
8. Anesthetic—inhalation (e.g., isoflurane) or injectable (e.g., ketamine [80–100 mg/kg] and xylazine [10–25.5 mg/kg]) (*see* **Note 2**).
9. Analgesic (e.g., carprofen 5 mg/kg).
10. Silk sutures.
11. Sterile surgical gloves.
12. Surgical gown.
13. Heat mat.
14. Chlorhexidine scrub.
15. Cotton tip applicators.
16. Skin glue.
17. Hair removal cream.

3 Methods

To minimize the risk of infection in the mouse, ensure that all procedures are performed under aseptic conditions. Sterilize all surgical instruments and consumables using an autoclave:

1. Weigh the mouse intended for surgery and record the weight.
2. Turn on the heat mat and prepare appropriate amount of anesthetic and analgesic based on the mouse's weight.
3. Administer anesthesia to the mouse using the selected method of anesthesia.
4. Administer the analgesic to the mouse before surgery, as they are more effective when given beforehand.
5. Place mouse in supine position and fix front paws to table using tape.
6. Remove fur on neck area using hair removal cream (*see* **Note 3**).
7. Sterilize the surgical area by applying chlorhexidine surgical scrub with a sterile cotton tip applicator, starting at the incision site and moving outward.
8. The investigator should clean hands with surgical scrub and then wear clean surgical gloves.
9. Under a dissecting microscope, make vertical midline 1 cm incision in the neck using sterile scissors.
10. Using retractors, widen the incision to expose surgical field (*see* **Note 3**).
11. Gently retract soft tissue over the trachea using sterile forceps.
12. Using sterile curved forceps, carefully separate the right common carotid artery from the carotid artery sheath and vagus nerve (*see* **Note 4**).
13. Loosely ligate the top and bottom of the right common carotid artery with two separate silk suture ties for easy manipulation of the artery (Fig. 1).
14. With the left hand, lift the top silk tie using a curved forceps to lift the right common carotid artery.
15. With the right hand, hold the bottom of a microcoil using a straight forceps, and place the microcoil on a slight angle directly underneath the right common carotid artery.
16. To manipulate the artery, move the top silk tie in circular motion to wrap the right common carotid artery around the coil, starting from the bottom end of the microcoil (*see* **Note 5**).

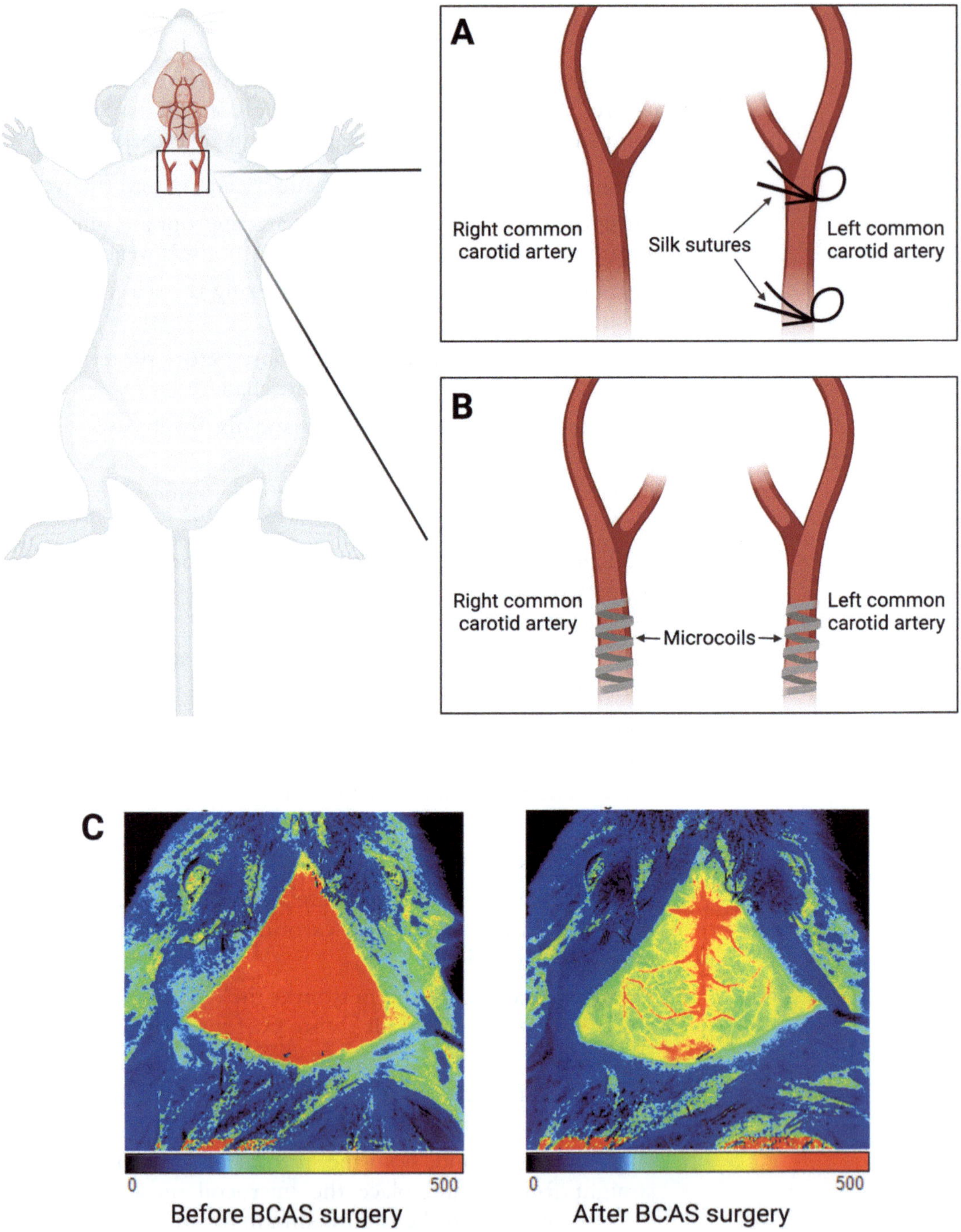

Fig. 1 Bilateral carotid artery stenosis surgery. (**a**) Silk sutures are loosely tied at the top and bottom of the common carotid artery for easy manipulation of the artery. (**b**) Both common carotid arteries are wrapped around microcoils to reduce cerebral blood flow. (**c**) There is an immediate drop in cerebral blood flow after BCAS surgery. The cerebral blood flow images were generated by the authors using laser speckle contrast imaging. (Figure created with BioRender.com)

17. With the left hand, carefully pick up the common carotid artery and coil using a curve forceps.
18. With the right hand, wrap the bottom end of the right common carotid artery around the remaining microcoil by using the bottom silk tie to manipulate the artery.
19. Blood flow should be reduced around the microcoil.
20. Remove silk ties. Blood flow should be restored in common carotid artery after removal of silk ties.
21. Repeat **steps 12–20** for the left common carotid artery.
22. Close incision site with skin glue.
23. Place mouse in heated recovery box until they regain consciousness.

Sham operated mice will undergo the same procedure except that no coils will be wrapped around the carotid arteries.

Postsurgical Monitoring

1. Follow institutional guidelines and monitor mice for any signs of pain, distress, and infection.
2. Administer 5 mg/kg/day carprofen for 2 days postsurgery as per institutional guidelines.

To ensure successful induction of cerebral hypoperfusion, cerebral blood flow can be measured before and immediately after surgery using techniques such as laser speckle contrast imaging or laser Doppler flowmetry. The duration of BCAS can vary. BCAS mice will begin to develop white matter lesions 15 days postsurgery [8]. Cognitive impairment has been observed 30 days postsurgery [9]. Mice have been subjected to BCAS for up to 8 months [7].

4 Conclusion

The BCAS mouse model induces chronic cerebral hypoperfusion which is a major contributor to the development of vascular dementia. Mice that undergo BCAS surgery develop the hallmarks of vascular dementia including cognitive impairment, neuroinflammation, and white matter lesions. The BCAS model can be used for acute or chronic studies of chronic cerebral hypoperfusion or vascular dementia in mice.

5 Notes

1. This is a recommendation for surgical instruments that can be used. Similar instruments may also be suitable.

2. The selection of anesthesia and analgesic should be made in compliance with institutional and other regulatory guidelines.
3. A hook-shaped tool can be used instead of retractor, e.g., bent needle on a syringe.
4. Be careful to not damage the vagus nerve as this can lead to overstimulation of the vagus nerve. Overstimulation can cause dyspnea and potentially death.
5. The common carotid arteries are flexible and can be easily wrapped around the microcoil. Be careful to not pierce the artery when handling it.

References

1. Cao Q et al (2020) The prevalence of dementia: a systematic review and meta-analysis. J Alzheimers Dis 73(3):1157–1166
2. O'Brien JT, Thomas A (2015) Vascular dementia. Lancet 386(10004):1698–1706
3. Du SQ et al (2017) Molecular mechanisms of vascular dementia: what can be learned from animal models of chronic cerebral hypoperfusion? Mol Neurobiol 54(5):3670–3682
4. Liang CS et al (2021) Mortality rates in Alzheimer's disease and non-Alzheimer's dementias: a systematic review and meta-analysis. Lancet Healthy Longev 2(8):e479–e488
5. Shibata M et al (2007) Selective impairment of working memory in a mouse model of chronic cerebral hypoperfusion. Stroke 38(10): 2826–2832
6. Shibata M et al (2004) White matter lesions and glial activation in a novel mouse model of chronic cerebral hypoperfusion. Stroke 35(11): 2598–2603
7. Nishio K et al (2010) A mouse model characterizing features of vascular dementia with hippocampal atrophy. Stroke 41(6):1278–1284
8. Rajeev V et al (2022) Intermittent fasting attenuates Hallmark vascular and neuronal pathologies in a mouse model of vascular cognitive impairment. Int J Biol Sci 18(16): 6052–6067
9. Selvaraji S et al (2022) Time-restricted feeding modulates the DNA methylation landscape, attenuates hallmark neuropathology and cognitive impairment in a mouse model of vascular dementia. Theranostics 12(7):3007–3023

Chapter 6

In Situ Hybridization to Characterize Neurulation and Midbrain–Hindbrain Boundary Formation in Zebrafish

Jarrad Fuller and Sebastian Dworkin

Abstract

Whole-mount in situ hybridization is able to harness the inherent advantages of zebrafish as a model organism for developmental biology, particularly when visualizing the formation of the neural tube, specifically at the level of the midbrain–hindbrain boundary. The size and transparency of developing zebrafish embryos allow for the visualization of neural markers in vivo along the length of the developing zebrafish central nervous system. In practice, this technique is useful for examining defects in neurulation and midbrain–hindbrain boundary formation that may arise following gene manipulation, for example, CRISPR mutagenesis. This method describes the process of embryo collection and preparation, RNA probe transcription, probe hybridization in vivo, as well as the process of probe detection and visualization.

Key words In situ hybridization, Zebrafish, Neurulation, Midbrain, Hindbrain, Development

1 Introduction

The zebrafish, *Danio rerio*, has rapidly emerged as an excellent experimental vertebrate model of development, particularly neurulation. Although the cellular mechanisms of neural tube formation differ somewhat between the fish and mammals, nonetheless, the genetic regulatory mechanisms of neurulation among zebrafish and mammals are remarkably well conserved. Therefore, experimental validation of zebrafish gene function provides substantial insight into genetic function during the brain development of higher mammals.

To this end, techniques that allow the visualization of dynamic changes in gene expression in vivo, such as whole-mount in situ hybridization (WISH), comprise a critical component in the

The original version of the chapter has been revised. A correction to this chapter can be found at https://doi.org/10.1007/978-1-0716-3585-8_19

Sebastian Dworkin (ed.), *Neurobiology: Methods and Protocols*, Methods in Molecular Biology, vol. 2746, https://doi.org/10.1007/978-1-0716-3585-8_6,

researchers' toolkit to better understand differential gene function following gene manipulation techniques, such as CRISPR mutagenesis or Morpholino oligonucleotide microinjection. This technique proves particularly practical and advantageous when applied to whole embryos of smaller organisms, such as zebrafish during their initial developmental stages. This method enables the detection of gene expression in a specific tissue sample by utilizing RNA riboprobes designed to hybridize to gene-of-interest mRNA, providing crucial information regarding spatiotemporal expression patterns.

In the context of neurulation, several markers have been very well described in the literature, and we have collated the expression patterns of these markers here. To visualize the neural tube, *sonic hedgehog* (*shh*) expression is an excellent marker, which also delineates the forebrain–midbrain boundary. *Orthodenticle homeobox 2* (*otx2*) is expressed strongly and specifically within the developing midbrain, and *paired box 2a* (*pax2a*) demarcates the optic stalk, otic vesicles, midbrain–hindbrain boundary (MHB), and hindbrain neurons [1]. Additionally, the *early growth response gene 2* (*krox20*) and *islet-1* (*isl1*) represent two examples of useful hindbrain markers [2, 3], detecting rhombomeres 3 and 5 and neuronal populations of the hindbrain, respectively. Anomalies in the expression of these markers following gene manipulation provide valuable insights into defects in the developing neural tube, mesencephalon, and rhombencephalon. To this end, we have previously used this WISH method to identify the role of CREB [4] and *Grainyhead-like* [5–8] transcription factors in various developmental processes, including neural and facial development.

This method encompasses embryo preparation, which involves dechorionation and fixation in paraformaldehyde. It also includes the generation of RNA probe templates through topoisomerase cloning, the synthesis of complementary RNA probes, and the hybridization of these probes to endogenous mRNA in whole embryos. Finally, detection of these probes is achieved through the application of a specific antibody and subsequent enzyme-based colorimetric development, allowing researchers to precisely determine the expression patterns of endogenous mRNA species at defined timepoints, to detect changes in both gene function and tissue morphogenesis.

2 Materials

2.1 Topoisomerase Cloning

1. Polymerase chain reaction (PCR) mix: 10 × buffer (100 mM Tris–HCl, pH 8.3, 500 mM KCl, 25 mM $MgCl_2$, 0.01% gelatin), 12.5 mM deoxynucleotide triphosphate (dNTPs), 100U Taq Polymerase, template-specific forward and reverse oligonucleotides (primers; 1 μM each primer), 250–1000 ng zebrafish cDNA.

2. TAE (Tris–acetic acid–ethylenediaminetetraacetic acid [EDTA]) 50 × stock: Dissolve 242 g Tris base in 500 mL water, and add 57.1 mL glacial acetic acid and 100 mL of 500 mM EDTA (pH 8.0) solution. Bring the final volume up to 1 L. Dilute to 1 × in water for working concentration.
3. 1% agarose gel: Dissolve 1 g agarose in 100 mL 1 × TAE. Boil for 60 s to dissolve, cool, and add 5 μL DNA-binding/visualization agent such as ethidium bromide or SybrSafe. Pour into gel mold and allow to set. Electrophorese DNA for 60 min at 100 V.
4. Salt solution: 200 mM NaCl, 10 mM $MgCl_2$.
5. Topoisomerase vector: 10 ng/μL plasmid DNA, 50% glycerol, 50 mM Tris–HCl (pH 7.4), 1 mM EDTA, 1 mM DTT, 0.1% Triton X-100, 100 μg/mL BSA (Invitrogen, MA) (*see* **Note 1**).
6. DH5α heat-shock competent *Escherichia coli* (*E. coli*) bacteria.
7. Heat source (e.g., heat block) for incubation at 42 °C.
8. SOC medium: 2% tryptone, 0.5% yeast extract, 10 mM NaCl, 2.5 mM KCl, 10 mM $MgCl_2$, 10 mM $MgSO_4$, 20 mM glucose.
9. Xgal/DMF: 40 mg/mL 5-bromo-4-chloro-3-indolyl β-D-galactopyranoside (X-gaL) in dimethylformamide (DMF).
10. Luria broth (LB): Combine 10 g of tryptone, 10 g of NaCl, and 5 g of yeast extract in 1 L of distilled water. Adjust the pH to 7.0 with 1 M NaOH and autoclave.
11. LB agar plates: Add 15 g agar to 1 L LB, stir well to dissolve, and autoclave. When the LB has cooled to ~55 °C (can be comfortably held in the hand), add 50 μg/mL ampicillin, and pour 10–20 mL into plastic bacterial culture plates.
12. Plasmid DNA extraction kit: Any commercial supplier.

2.2 Embryo Preparation

1. Phosphate buffered saline (PBS): For a 10× solution, add 80 g NaCl, 2 g KCl, 14.4 g Na_2HPO_4, add 800 mL of distilled water, adjust to pH 7.4, and fill to 1 L. For 1× PBS, add 10 mL of 10× PBS to 90 mL of distilled water.
2. 4% paraformaldehyde: 40 g of paraformaldehyde, 800 mL 1× PBS. Place on heating block at 60 °C but do not boil. Stir until dissolved and solution appears clear. Adjust to pH 7 with 5 M NaOH. Filter to remove particulates and adjust to 1 L with 1× PBS. Store as 10 mL aliquots at −20 °C.
3. E3 embryo medium: Add 34.8 g of NaCl, 1.6 g of KCl, 5.8 g $CaCl_2.2H_2O$ to 2 L of water. Adjust pH to 7.2 and autoclave. Dilute 16.5 mL of this solution in 1 L of distilled water for a 1× preparation.
4. Pronase: 100 mg of Pronase in 10 mL of distilled water (10 mg/mL), dilute 1 mL of this stock in 9 mL of E3 embryo medium for 1 mg/mL solution.

2.3 RNA Probe Transcription

1. Restriction endonucleases and buffers: Any commercial supplier, as required based on vector and insert restriction sites.
2. RNA polymerase: Any commercial supplier, Sp6/T3/T7 RNA polymerases, as required.
3. Digoxigenin (DIG) RNA labelling mix (Roche).
4. 0.1% Tween 20 in PBS (PBST): Add 100 μL of Tween 20 solution to 100 mL of 1× PBS.
5. Saline sodium citrate (SSC): Add 88.2 g Na_3-citrate and 175.3 g NaCl and 800 mL of distilled water. Adjust the pH to 7.0 with 1 M HCl. Fill to 1 L with distilled water.
6. 2×, 1×, and 0.2× saline sodium citrate–Tween (SSCT): 20 × SSC diluted to desired concentration, supplemented with 0.1% Tween 20.
7. Proteinase K: Dissolve powdered stock in distilled water at a final concentration of 20 mg/mL. Store frozen in 1 mL aliquots.
8. Prehybridization buffer (HYB−): 500 mL formamide, 250 mL 20× SSC, 1 mL Tween 20, 250 mL distilled water.
9. Hybridization buffer (HYB+): 499.5ML HYB− buffer, 5 mg/mL Torula RNA powder, 500 μL of Tween 20.
10. Riboprobe solution: RNA probe (50 ngl/μL) in HYB+ solution.
11. 2 × SSCT/50%formamide: 50 mL of 20 × SSC, 250 mL formamide, 200 mL of distilled water, 500 μL of Tween 20.
12. RNase solutions: RNASE A (20 mg/mL stock) RNASE T1 (100,000 u/mL stock).
13. Blocking solution: 5% fetal calf serum, 2 mg/mL bovine serum albumin, 1% dimethyl sulfoxide in PBST.
14. Anti-DIG Fab alkaline phosphatase antibody.
15. Stain buffer: 100 mM Tris pH 9.5, 50 mM $MgCl_2$, 100 mM NaCl, 0.1% Tween 20, 1 mM levamisole, add distilled water to final volume.
16. Stain substrate: 0.33 mg/mL nitro blue tetrazolium salt (NBT) and 0.17 mg/mL 5-bromo-3-chloro-3-indolyl phosphate (BCIP), diluted in stain buffer.
17. Netwell inserts, 15 mm diameter.
18. 6-well culture plates.
19. Aluminum foil.

3 Methods

3.1 PCR Amplification and Cloning Preparation

1. Design primers for your desired probe template product (*see* **Note 2**) and amplify using a mix consisting of 5 μL of 10× buffer, 0.5 μL of dNTP's (50 mM), 1 μM of primers, 100 ng of DNA template. Add water to a final volume of 49 μL before adding 1 μL Taq Polymerase.
2. The above reaction mix was amplified using a denaturation step of 94 °C for 2 min and 30 s, 35 cycles of 94 °C for 30 s, 65 °C for 1 min and 72 °C for 2 min and 30 s, followed by a final extension step of 72 °C for 7 min.
3. If desired, run 5 μL of PCR product using standard agarose gel electrophoresis (and a 1% agarose gel) to ensure the product is of appropriate size and that there is only a single band (*see* **Note 3**).
4. Prepare a "plasmid map" comprising the predicted sequence of the insert within the topoisomerase vector. As the insert may ultimately enter the vector in either 5′-3′ or 3′-5′ orientations, a plasmid map for each orientation should be prepared (Fig. 1).

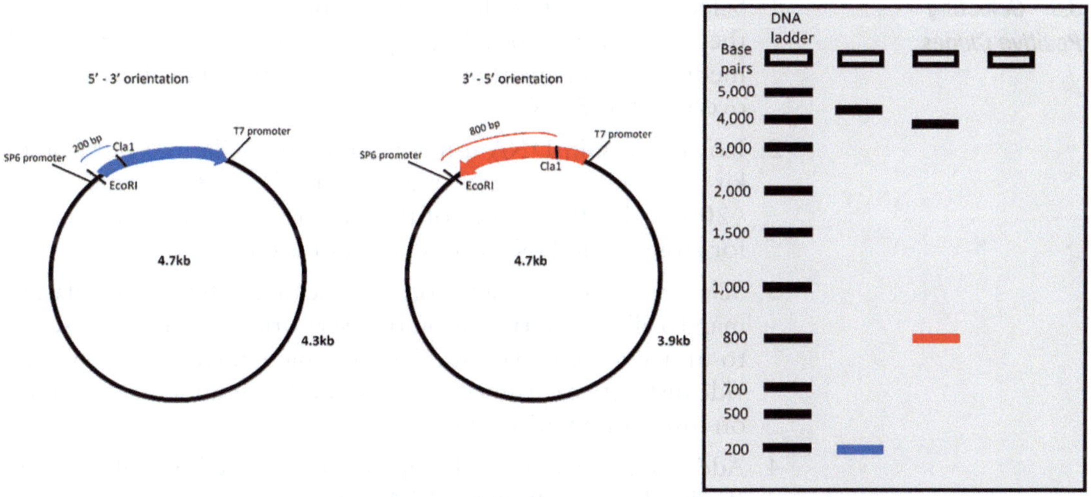

Fig. 1 An example of a plasmid map using restriction enzymes to determine the orientation of an 800 base pair DNA fragment within a topo-TA vector. Restriction enzyme sites—in this instance, EcoRI present on the topo-TA plasmid and Cla1 present on the insert allow the insert to be cut following cloning. The fragments produced following digestion with restriction enzymes will produce fragments of differing sizes based upon the orientation of the insert within the host plasmid. In this instance, a 5′–3′ oriented fragment produced a 200 bp band following agarose gel electrophoresis, while a 3′–5′ oriented fragment produced an 800 bp band. In this way, the desired orientation of the insert can be determined following cloning

3.2 Topoisomerase Cloning

1. To begin the cloning reaction, assemble a mix of 3 μL of PCR product, 1 μL of salt solution, 1 μL of topoisomerase vector and water to a final volume of 6 μL.
2. Once assembled, mix the solution for 5 min, before placing the reaction on ice for 5 min.
3. While the cloning reaction is incubating, spread LB Agar plates with X-gal/DMF, allow to dry for 10 min, and pre-warm to 37 °C.
4. Add 2 μL of the cloning reaction into a vial of heat-shock competent *E.coli*, mix and incubate on ice for 20 min.
5. Place the cells at 42 °C for 30 s (without shaking) to heat shock cells, before returning to ice for 2 min.
6. Add 250 μL of SOC medium, cap the tube and shake horizontally for 1 h at 37 °C.
7. Take 40 μL of each transformation and spread on two pre-warmed LB/ampicillin plates from **step 1**.
8. Add 20 μL of SOC medium to each plate so that the cells spread evenly across the plates.
9. Incubate plates at 37 °C for a minimum of 16 h.

3.3 Selecting Positive Clones

1. Select 6–12 white colonies by scraping individual colonies off the plate with a 20–200 μL pipette tip and add to 5 mL LB medium supplemented with 50 μg/mL ampicillin. Culture overnight at 37 °C, with shaking.
2. Isolate plasmid DNA using a commercial plasmid purification kit according to the manufacturer's instructions. Leave ~200 μL in the bottom of the tube to use as a starter culture for larger-scale DNA extraction if desired.
3. Select appropriate restriction enzymes according to the plasmid map to allow determination of insert orientation. It is essential to utilize a combination of compatible enzymes that will provide distinctive (diagnostic) sizes of DNA fragments depending on insert orientation (Fig. 1).
4. Add 2 μg plasmid, 2 μL appropriate restriction buffers, 2 μg Bovine Serum Albumin (BSA) and 10–20 units of restriction endonuclease, in a final volume of 20 μL in H_2O.
5. Incubate at a temperature appropriate for each enzyme (according to the manufacturer's specifications) for 2–3 h.
6. Electrophorese 10 μL of this reaction on a 1% agarose gel to determine insert orientation.
7. If desired, use the 200 μL remnant culture (in **step 2** above) of a positive clone identified to inoculate 50-250 mL of LB-Ampicillin for larger-scale plasmid propagation and extraction.

3.4 Riboprobe Synthesis

1. Linearize 1-2 μg plasmids with the appropriate restriction enzyme as outlined in Subheading 3.3, **step 3**.
2. Set aside 3 μL of the digested plasmid product and run it on a 1% agarose gel in order to check that the plasmid was linearized efficiently.
3. Precipitate DNA using 9.5 μL of 3 M NaOAc in 95 μL of sample.
4. Add 190 μL of absolute ethanol per 95 μL of sample and centrifuge at 14,000 × *g* for 30 min at 4 °C.
5. Air dry the resulting pellet and wash in 70% ethanol, centrifuge at 14,000 × *g* for 5 min at 4 °C and air dry again.
6. Resuspend the pellet in 50 μL of distilled water, dilute to a concentration of approximately 100–150 ng/μL.
7. To transcribe the cRNA riboprobes, add 2 μL of 10× transcription buffer, 2 μL of Digoxigenin (DIG) RNA labelling mix, 1 μL RNAse inhibitor, 1 μg linearized DNA template, 2 μL RNA polymerase (T7/T3/SP6) with nuclease free water to a final volume of 20 μL, on ice.
8. Incubate the transcription reaction at 37 °C for 3 h, add 1 μL of RNase-free DNase to digest the plasmid DNA template and incubate the solution for an additional 30 min at 37 °C.
9. Next, precipitate the riboprobe in 4 μL of 4 M LiCl and 75 μL 100% ethanol in order to remove unincorporated nucleotides.
10. Leave the riboprobe solution at −70 °C for 2 h and then centrifuge at 14,000 × *g* for 30 min at 4 °C.
11. Wash the resulting pellet in 70% ethanol and centrifuge at 14,000 × *g* at 4 °C for 5 min, air dry, and resuspend in 50 μL sterile water.
12. If desired, take a 5 μL of riboprobe solution and run it on a 1% TAE agarose gel to confirm successful transcription and size of the cRNA probe.
13. To prepare a working concentration RNA probe, dilute an aliquot of the stock to 1 ng/μL in HYB+ solution (you will use ~500 μL HYB+ per reaction). This diluted probe can be stored at −20 °C for up to 6 months.

Embryo Preparation

1. Dechorionate embryos under a dissecting microscope using forceps. Carefully peel away the chorion without damaging the embryo inside.
2. Fix dechorionated embryos overnight in 4% PFA/PBS.
3. Rinse fixed embryos in PBST 3 × 5 min.

4. Dehydrate embryos in sequential washes of methanol/PBST comprising 25%, 50%, 75% and 2 × 100% methanol, for a duration of 10 min per step, before placing the embryos in methanol overnight at −20 °C (*see* **Note 4**).
5. Warm the embryos to room temperature and rehydrate in sequential washes of methanol/PBST comprising 75%, 50% and 25% methanol, for a duration of 10 min per step, followed by 2 × 5 min washes in PBST. (*see* **Note 5**).
6. Digest the embryos in ~10 mL of Proteinase K at room temperature; the length of digestion is dependent on the age of the embryo:
   (a) <24 hpf – no digestion.
   (b) 24 hpf embryos – 5 min in 10 μg/mL proteinase K.
   (c) 36 hpf – 10 min in 10 μg/mL proteinase K.
   (d) >36 hpf – 20–30 min in 20 μg/mL proteinase K.
7. Rinse the embryos in PBST.
8. Fix embryos in 4% PFA/PBS for 20 min.
9. Rinse embryos 4 × 5 min in PBST to remove fixative.

Hybridization

1. Transfer embryos from PBST to 500 μL of pre-warmed HYB- solution and incubate at 70 °C for 15 min (*see* **Note 6**).
2. Transfer embryos to pre-warmed HYB+ solution and incubate for 4 h at 70 °C.
3. Transfer embryos to pre-warmed riboprobe solution and incubate overnight at 70 °C (*see* **Note** 7).

Probe Removal

1. Remove embryos from riboprobe solution and place in net-wells in 50% Formamide/2 × SSCT at 70 °C for the following series of washes (*see* **Note 8**).
2. Incubate embryos for 30 min at 70 °C in formamide/ 2 × SSCT, followed by 3 × 10 min washes in 2 × SSCT at 37 °C and 1 × 10 min wash in PBST at 37 °C.
3. Incubate embryos in 20 μg/mL RNaseA in PBST for 30 min at 37 °C.
4. Incubate embryos in 2 × SSCT for 10 min at 37 °C (*see* **Note 9**).
5. Incubate embryos at 70 °C for 1 h in pre-warmed 50% formamide/2 × SSCT.
6. Incubate embryos for 15 min in 2 × SSCT at 70 °C, and then 2 × 15 min in 0.2× SSCT at 65 °C.

Detection

1. Wash embryos 3 × 5 min in PBST, then incubate in blocking solution for 2 h in the dark (as DMSO is light sensitive).
2. Incubate embryos overnight at 4 °C in 1:5000 anti-DIG Fab-AP antibody fragments in blocking solution. It is recommended that embryos be moved to 24-well plates to minimize the amount of antibody needed and incubated (overnight) in the dark.

Colour Development

1. Transfer embryos back into Netwells in 6 well plates, rinse embryos in PBST to remove antibody/blocking solution, then wash 4 × 60 min in PBST.
2. Remove embryos from PBST and wash twice for 8 min in staining buffer.
3. Transfer embryos from netwells to 24 well plates with staining buffer and staining substrate added and leave to develop in the dark without agitation.
4. Once the staining has reached a satisfactory level for imaging (or when background staining is present), stop the reaction by washing twice in PBST for 5 min.
5. Embryos may be stored for up to 2 weeks in 4% PFA.
6. Common applications when imaging the expression patterns of well-validated neural probes (Fig. 2) can determine neural regions that may be absent or aberrantly patterned following gene manipulation (Fig. 3) or may be used to identify changes in brain and neural tube morphogenesis (Fig. 4).

4 Notes

1. We routinely used the pCRII-TOPO vector, as it is a linearized plasmid harbouring single 3′-thymidine (T) overhangs alongside topoisomerase I covalently bound to the vector; this allows TA cloning when combined with a PCR-product. It is important to ensure that a non-proofreading *Taq* Polymerase enzyme is used for the PCR reaction, as this will add a single adenosine (A) nucleotide to the 3′-end. This is required to facilitate ligation into the vector.
2. In order to ensure specificity of subsequent probe and allow for high-stringency binding conditions, we routinely design our amplicons to be between 400 and 800 base pairs. It is also helpful if they are comprised of not more than 55% guanine and cytosine nucleotides (i.e., not overly "GC-rich").

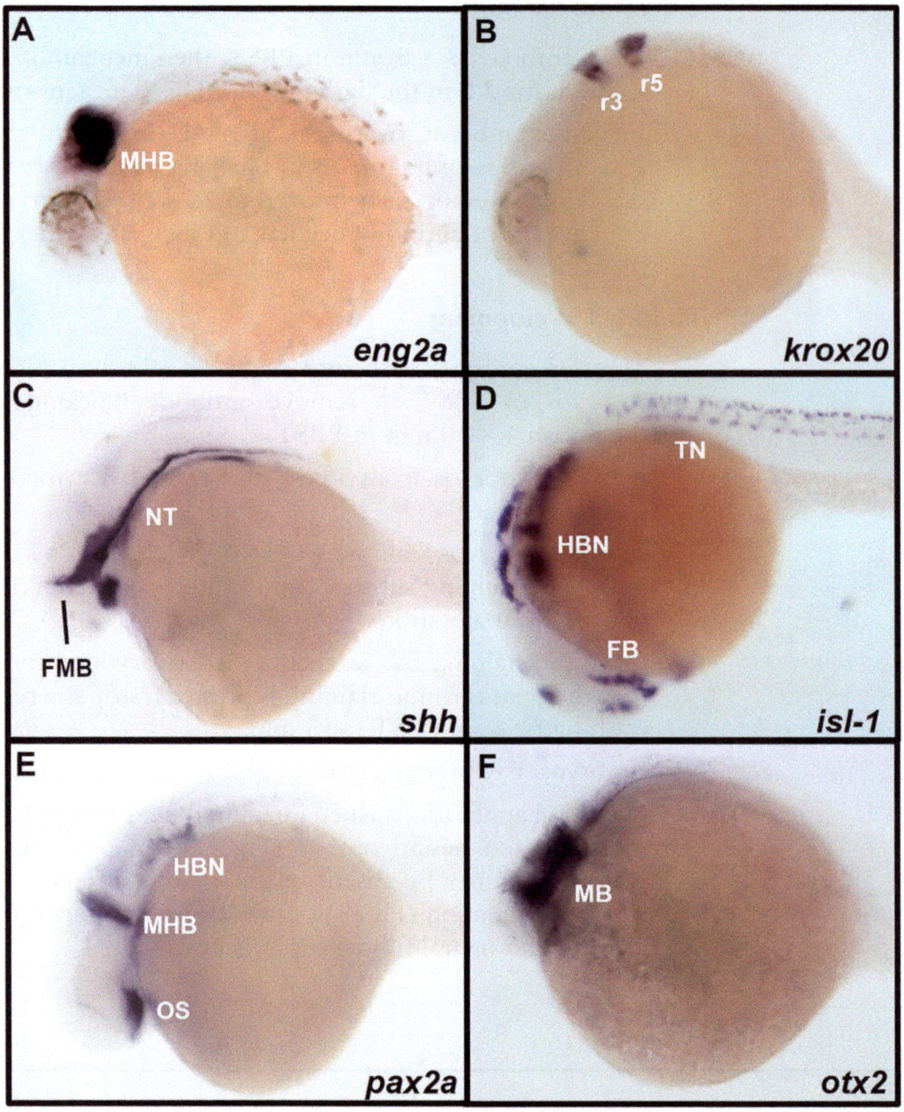

Fig. 2 Expression patterns of commonly used probes to study neurulation in zebrafish. (**a**) Engrailed2a (*eng2a*) demarcates the midbrain–hindbrain boundary (MHB). (**b**) Early growth response gene 2 (*krox20*) is expressed in rhombomeres 3 and 5 (r3/r5) in the hindbrain. (**c**) Sonic hedgehog (*shh*) is expressed in the frontal–midbrain boundary (FMB) and the neural tube (NT). (**d**) Islet-1 (isl-1) is expressed within the forebrain (FB), hindbrain neurons (HBN), and also neurons in the trunk (TN). (**e**) Paired box 2a (*pax2a*) is expressed in the optic stalk (OS), MHB, and HBN. (**f**) Orthodenticle homeobox 2 (*otx2*) is expressed throughout the midbrain (MB). Together, these probes can be used to identify defects in cellular patterning and organ formation

3. This step is highly recommended to validate primer fidelity.
4. At this point, embryos may be stored at −20 °C for up to 3 months.
5. For washing and dehydrating/hydrating steps, it is easiest to use 6-well plates. Using netwells will minimize damage to the embryos, which is critically important to keeping the embryos intact throughout the experiment.

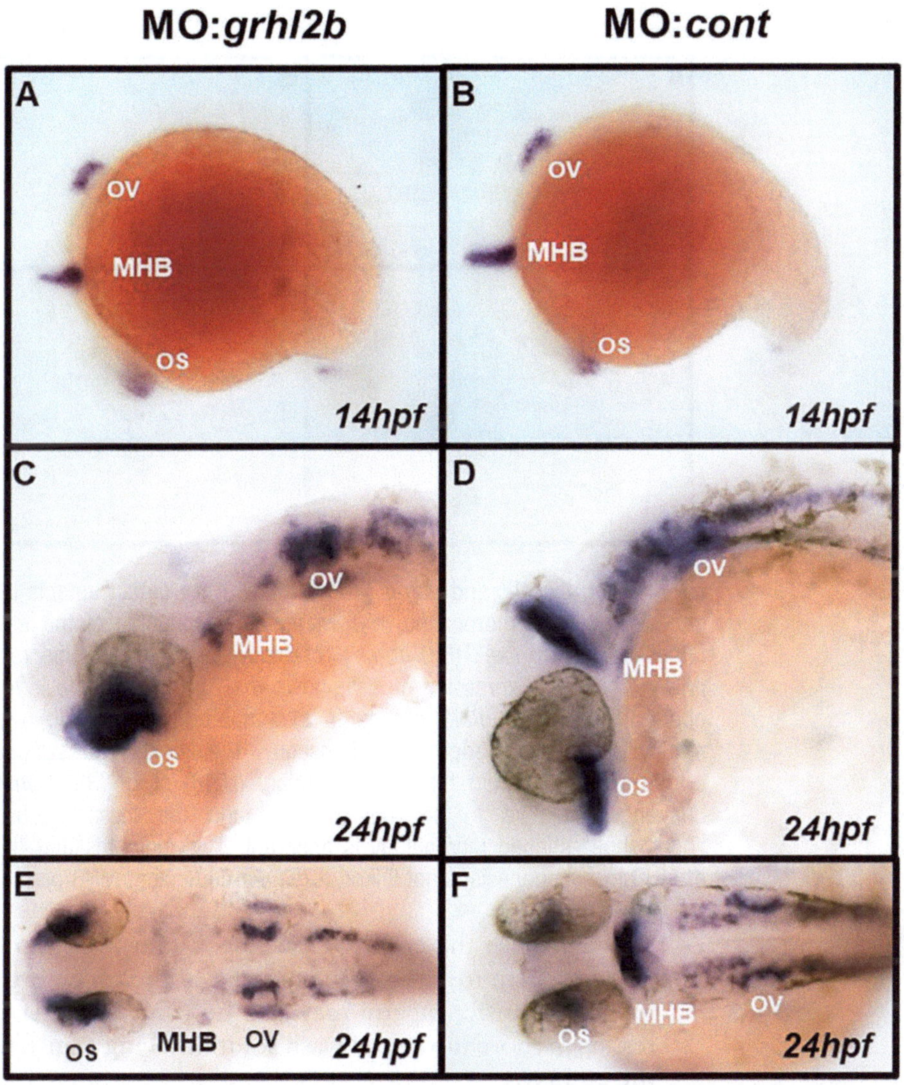

Fig. 3 Using the *pax2a* probe for WISH to detect defects in cellular patterning following morpholino (MO)-mediated knockdown of the transcription factor *grainyhead-like 2b* (MO/*grhl2b*; **a**, **c**, **e**) relative to MO/control injected embryos (MO/*cont*; **b**, **d**, **f**). At 14 h post-fertilization (hpf; **a**, **b**), there is little difference in *pax2a* expression in either the otic vesicle (OV), midbrain–hindbrain boundary (MHB), or optic stalk (OS). However, by 24hpf, *pax2a* expression is completely absent in the MHB region of MO/*grhl2b*-injected embryos (**c**, lateral; **e**, dorsal) relative to the strong expression seen in MO/*cont* injected embryos (**d**, lateral; **f**, dorsal). The continued strong expression of *pax2a* in the OS and OV of MO/*grhl2b* injected embryos serves as an excellent internal positive control to ensure that the WISH process has worked optimally and indicates that these regions are not aberrantly patterned in MO/*grhl2b* embryos

6. Embryos should be moved from a 6-well plate to a flat-bottom tube to minimize the amount of HYB solution used. The easiest way to transfer embryos is to cut the bottom off a standard 100 mL pipette tip with scissors so that the gauge of the tip is larger, and carefully draw up the embryos from the

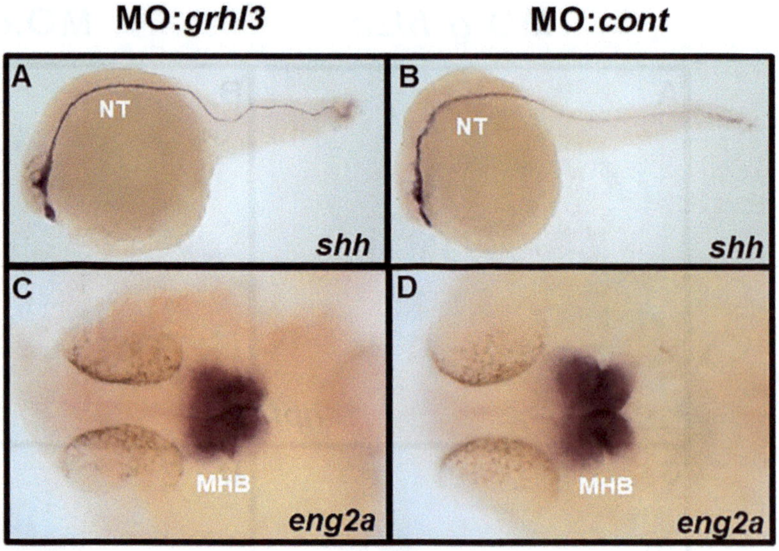

Fig. 4 Using the *shh* and *eng2a* probes for WISH to detect defects in morphogenesis following morpholino (MO)-mediated knockdown of the transcription factor *grainyhead-like 3* (MO/*grhl3*; **a**, **c**) relative to MO/control injected embryos (MO/*cont*; **b**, **d**). Expression of *shh* shows that the neural tube is kinked and aberrantly shaped in *grhl3*-morphants (**a**) compared to controls (**b**); a defect not readily visible under brightfield microscopy. The strong expression pattern of *eng2a* in the midbrain–hindbrain boundary (MHB) of *grhl3*-morphants (**c**) is unchanged relative to controls (**d**), although the loss of characteristic "horseshoe"-shaped morphology is apparent. This indicates that the process of MHB morphogenesis is not linked to *eng2a*-dependent MHB patterning

netwell. Once the embryos have been drawn up into the tip, wait for them to settle at the bottom of the tip before ejecting them. A transfer pipette may also be used, but this generally results in more unwanted wash solution being transferred with the embryos.

7. It is especially important to preheat the riboprobe solution(s) at 68 °C for at least 10 min, so that secondary structure is removed from the probe before addition of embryos.
8. Riboprobe solution can be kept for reuse between 3 and 8 times without noticeable loss of signal.
9. **Steps 2–4** in this section are recommended only for probes that present with suboptimal binding or significant background staining. For the majority of well-designed, stable, intact probes, this step is not necessary. If a probe consistently yields substantial background, it is recommended to either transcribe a fresh RNA probe (as in Subheading 3.4) or redesign an entirely new probe template from a distinct part of the coding sequence (as in Subheading 3.1–3.3).

10. The development of staining is quite variable—in some instances, staining may be observed within 2 h; however, 6+ h incubations may be necessary to observe staining.

References

1. Dworkin S, Darido C, Georgy SR, Wilanowski T, Srivastava S, Ellett F et al (2012) Midbrain-hindbrain boundary patterning and morphogenesis are regulated by diverse grainy head-like 2-dependent pathways. Development 139(3):525–536. https://doi.org/10.1242/dev.066522
2. Picker A, Scholpp S, Böhli H, Takeda H, Brand M (2002) A novel positive transcriptional feedback loop in midbrain-hindbrain boundary development is revealed through analysis of the zebrafish pax2.1 promoter in transgenic lines. Development 129(13):3227–3239. https://doi.org/10.1242/dev.129.13.3227. PMID: 12070097.
3. Giudicelli F, Taillebourg E, Charnay P, Gilardi-Hebenstreit P (2001) Krox-20 patterns the hindbrain through both cell-autonomous and non cell-autonomous mechanisms. Genes Dev 15(5):567–580. https://doi.org/10.1101/gad.189801. PMID: 11238377; PMCID: PMC312642
4. Dworkin S, Heath JK, deJong-Curtain TA, Hogan BM, Lieschke GJ, Malaterre J et al (2007) CREB activity modulates neural cell proliferation, midbrain-hindbrain organization and patterning in zebrafish. Dev Biol 307(1): 127–141. https://doi.org/10.1016/j.ydbio.2007.04.026
5. Dworkin S, Simkin J, Darido C, Partridge DD, Georgy SR, Caddy J et al (2014) Grainyhead-like 3 regulation of endothelin-1 in the pharyngeal endoderm is critical for growth and development of the craniofacial skeleton. Mech Dev 133:77–90. https://doi.org/10.1016/j.mod.2014.05.005
6. Gasperoni J, Fuller J, Darido C, Wilanowski T, Dworkin S (2022) Grainyhead-like (Grhl) target genes in development and cancer. Int J Mol Sci 23:2735. https://doi.org/10.3390/ijms23052735
7. Mathiyalagan N, Dworkin S (2022) Whole-mount in-situ hybridization (WISH) in zebrafish embryos to analyze craniofacial development. Methods Mol Biol 2403:19–32. https://doi.org/10.1007/978-1-0716-1847-9_2. PMID: 34913113
8. Miles LB, Darido C, Kaslin J, Heath JK, Jane SM, Dworkin S (2017) Mis-expression of grainyhead-like transcription factors in zebrafish leads to defects in enveloping layer (EVL) integrity, cellular morphogenesis and axial extension. Sci Rep 7(1):17607. https://doi.org/10.1038/s41598-017-17898-7

Chapter 7

Administering a Behavioral Test Battery in Rodents

Emily J. Jaehne, Michelle Corrone, and Maarten van den Buuse

Abstract

Although animal models cannot broadly represent uniquely human psychiatric or psychological syndromes such as anxiety, depression, or schizophrenia, behavioral testing in rodents can be extremely helpful to investigate specific disease aspects and symptoms. Animal behavioral test batteries allow researchers to reveal specific behavioral changes in genetically modified mice or following targeted treatments or in response to environmental interventions. Examples of types of behaviors that can be combined in a test battery include anxiety-like behavior, learning and memory, depression-relevant behavior, social interaction, and locomotor hyperactivity. Here, we describe several commonly used and relatively simple behavioral tests which can be combined in the same cohort of animals.

Key words Behavior, Phenotyping, Test battery, Learning, Memory, Cognition, Anxiety, Depression, Social interaction, Fear, Locomotor activity

1 Introduction

Although behavioral animal models cannot broadly represent uniquely human psychiatric or psychological syndromes such as anxiety, depression, or schizophrenia, behavioral testing in rodents can be extremely helpful to investigate specific disease aspects and symptoms as well as brain mechanisms involved [1, 2]. Animal models also allow researchers to test for specific behavioral or physiological changes in genetically modified mice or following targeted treatments or environmental interventions [1–4].

Examples of types of behaviors that can be assessed include fear and anxiety-like behavior, learning and memory, depression-relevant behavior, social interaction, and locomotor hyperactivity. Anxiety-like behaviors can be studied using simple tests such as elevated plus maze (EPM) and open field (OF). Mice which spend more time in the open arms of the EPM or the center of the OF can be considered to have a lesser anxiety-like phenotype [5–7]. Several models to test for cognitive impairment [8], including the Y-maze and novel object recognition test (NORT), rely on

Sebastian Dworkin (ed.), *Neurobiology: Methods and Protocols*, Methods in Molecular Biology, vol. 2746,
https://doi.org/10.1007/978-1-0716-3585-8_7,

rodents' innate preference to explore novel locations or objects [9–11]. The forced swim test (FST) was developed to assess the efficaciousness of antidepressant drugs to increase active coping strategies (such as swimming and climbing) and decrease passive coping behaviors (immobility) adopted by rodents [12, 13]. Social interaction can be studied using the three-chamber social approach task [14, 15]. Preference for interacting with a "stranger" mouse over an empty cage can be used as a measure of sociability, [16] while preference for interaction with a novel stranger mouse over a familiar mouse gives a measure of preference for social novelty or memory of the familiar stranger [17]. The addition of specific challenge drug treatments adds a degree of pharmacological specificity, informing on the involvement of selected neurotransmitter systems involved in the behavioral changes observed. This is particularly important for psychosis-like behaviors such as locomotor hyperactivity, which ultimately can aid to model the positive symptoms of schizophrenia [4] or related conditions.

It is common to conduct two or more of these behavioral tests in the same animals in a behavioral battery [18–21]. While there are clear ethical and economic advantages to performing behavioral batteries, as fewer mice will be used than if each test if conducted in a separate cohort, there are also advantages in interpretation of results [22]. For example, if a mouse is shown to display anxiety-like behavior in the EPM or OF and also shows low preference for the novel arm in the Y-maze or the novel object in the NORT, there is a chance that their increased anxious phenotype may have led to what would otherwise appear to be impaired recognition memory. Another advantage of test batteries is that statistical correlation between data from different tests can be done if these are performed in the same animals. However, it is also important to take into account that there are possible limitations to using behavioral batteries, such as responses in one test could be influenced by experiences in a previous test [22]. To limit this, it is common to perform tests that can be considered the least stressful first and the presumably most stressful tests (e.g., those including exposure to tones, mild foot shocks, or drug administration) last [22]. It is also important to allow a number of days (minimum two, recommended three or four) between different behavioral tests to reduce possible cumulative stress of a series of behavioral tests in a test battery. Figure 1 shows a number of possible combinations of behavioral tests in mice. The choice of which tests are included depends on the research focus of the project. However, it is recommended to include a broad range of tests to be able to draw conclusions on the specificity of any behavioral changes observed [3, 8, 23].

Here, we describe several commonly used behavioral tests which can be combined in the same cohort of animals with appropriate intervals between tests. We have previously shown examples

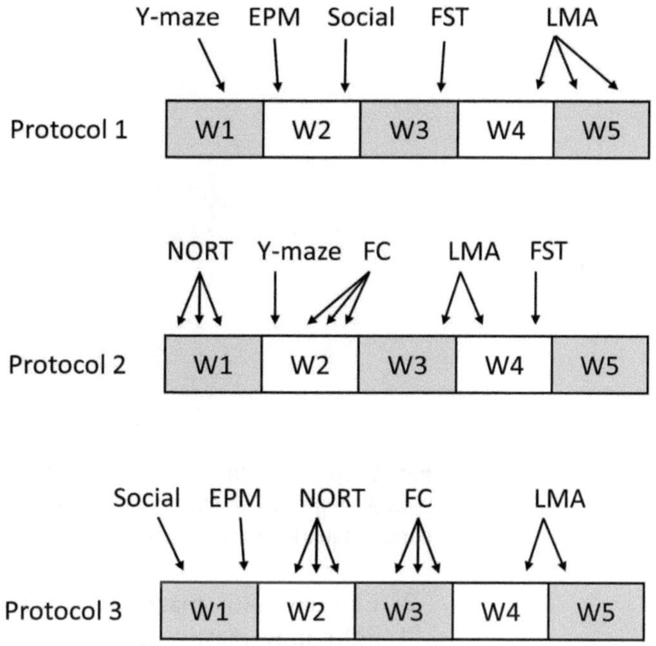

Fig. 1 Examples of three possible test batteries in rodents. Horizontal bars indicate weeks (W1, W2, etc.). Different tests can be combined in the same cohort of mice, with a minimum of two days interval. EPM = elevated plus maze; Social = social interaction and social recognition memory; FST = forced swim test; LMA = locomotor activity; NORT = novel object recognition test; FC = fear conditioning

of different combinations of these tests in different cohorts of mice [18, 19, 21, 23]. We will show details of the individual tests as they have been used in our laboratory for many years, but slight variation in sizes, materials, and protocol details are possible. We will focus on mice, but similar tests are available for rats, usually with analogues methodology but larger apparatuses than the descriptions below.

2 Materials

2.1 General

1. Experimental and control mice. In the case of social interaction behavior, young adult "stranger" mice of the same strain/sex as the test mice but not littermates.
2. 80% ethanol in spray bottle and paper towel.

2.2 Elevated Plus Maze

1. Video tracking software and hardware such as EthoVision (Noldus, The Netherlands) or similar or a video camera for recording behavior for offline scoring.

Fig. 2 Top view of the elevated plus maze: Two open and two closed arms, each with a length of 32 cm and width of 5 cm, 30 cm above the ground, with a central square section between arms and walls 20 cm high on the closed arms

2. Elevated plus maze: Elevated plus-shaped platform made of light gray plastic, including two open and two closed arms, each with a length of 32 cm and width of 5 cm, 30 cm above the ground, with a central square section between arms and walls 20 cm high on the closed arms (Fig. 2). To reduce the chance of the mouse falling off the open arms, a wall of approximately 0.5 cm high can be added on these arms. Additional white floor inserts can be added to assist with video tracking of dark-colored mice.

2.3 Open Field

1. Video tracking software and hardware such as EthoVision (Noldus, The Netherlands) or similar or a video camera for recording behavior for offline scoring.
2. Open field: Nontransparent 40 × 40 cm arena with walls 30 cm high (Fig. 3). Made of light gray plastic. Additional white floor inserts can be added to assist with tracking of dark-colored mice.

2.4 Novel Object Recognition Test

1. Video tracking software and hardware such as EthoVision (Noldus, The Netherlands) or similar or a video camera for recording behavior for offline scoring.
2. Open field: Nontransparent 40 × 40 cm arena with walls 30 cm high (Fig. 3). Made of light gray plastic. Additional white floor inserts can be added to assist with tracking of dark-colored mice.
3. Assorted pairs of objects of different colors and textures. Approximately 4–6 cm diameter and 5–8 cm tall (Fig. 3).

Fig. 3 Open field: Nontransparent 40 × 40 cm arena with walls 30 cm high. Pictured inside are two nonidentical objects as used for the novel object recognition test

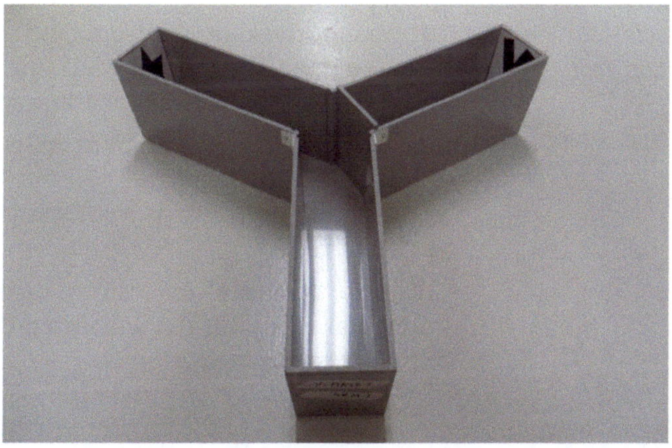

Fig. 4 Y-maze: Three arms (start arm and two test arms), each 32 cm long and 10 cm wide with walls 15 cm high. The two test arms have different black and white symbols on either end wall

2.5 Y-Maze Spatial Recognition Memory

1. Video tracking software and hardware such as EthoVision (Noldus, The Netherlands) or similar or a video camera for recording behavior for offline scoring.
2. Y-maze: Y-shaped apparatus made of light gray plastic with three arms (start arm and two test arms), each 32 cm long and 10 cm wide with walls 15 cm high (Fig. 4). The two test arms have different black and white symbols on either end wall.
3. Sawdust bedding material.

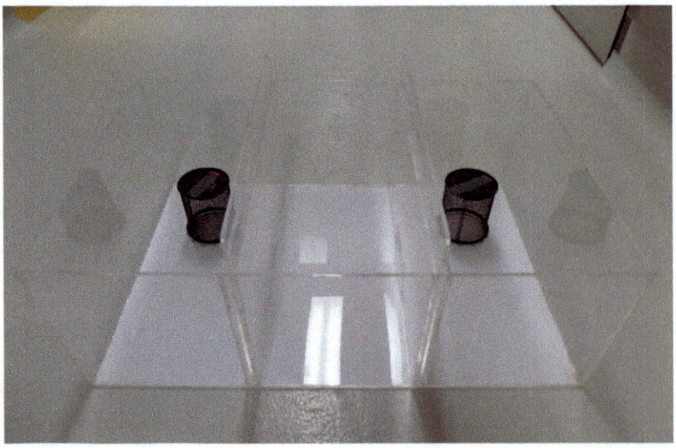

Fig. 5 Three-chamber sociability/social memory test: rectangular three-chambered enclosure, 43 cm × 64 cm, with transparent walls 23 cm high, with removable barriers between chambers. Pictured inside are two "stranger" enclosures

2.6 Social Interaction Test

1. Video tracking software and hardware such as EthoVision (Noldus, The Netherlands) or similar or a video camera for recording behavior for offline scoring.
2. Three-chamber sociability test apparatus. Rectangular three-chambered enclosure, 43 cm × 64 cm, with transparent walls 23 cm high, with removable barriers between chambers (Fig. 5).
3. Two "stranger" enclosures, diameter 9 cm, height 10 cm, with metal grid walls to allow interaction between mice but prevent aggressive interactions.

2.7 Forced Swim Test

1. Video camera for recording behavior for offline scoring.
2. Stopwatch.
3. 2 L transparent cylinders or beakers.
4. Thermometer.
5. Sink access.
6. Heater and heat pads.
7. Towels.
8. Behavioral scoring software such as Kinoscope [24].

2.8 Fear Conditioning Test

1. Automated setup such as Med Associates NIR Video Fear Conditioning (VFC) System which includes VideoFreeze Video Fear Conditioning software. See https://med-associates.com/product-category/learning-memory-attention/fear-conditioning/ for images and technical details of equipment.

2. Diluted peppermint essence and water in separate spray bottles.
3. Sawdust bedding material.
4. Wall inserts.

2.9 Locomotor (Hyper)Activity

1. Automated setup such as Med Associates Seamless Open Field Package including Infrared beams for tracking, and activity monitor software. See https://med-associates.com/product-category/anxiety-depression/open-field-activity/ for images and technical details of equipment. Similar equipment is available from other suppliers.

3 Methods

3.1 General

1. Allow mice to habituate to the animal holding area and behavioral testing facility for at least 7 days prior to the start of any experiments.
2. Set up relevant software for each behavioral test and confirm ready to start recording data prior to placing mouse in apparatus (*see* **Note 1**).
3. Press start on camera or tracking software immediately after placing mouse in the apparatus, and remove mouse immediately following preset analysis period on tracking software or press stop on camera as soon as mouse is removed (*see* **Note 1**).
4. Ensure all equipment is cleaned with 80% ethanol solution or other specified solution, prior to testing of the first mouse and in-between subsequent mice. Allow apparatus to dry before testing mouse.
5. At the end of any behavioral test, inspect each mouse for any health concerns (*see* **Note 2**) before returning to their home cage. Return all mice to their housing room at the end of the daily testing period.

3.2 Elevated Plus Maze

1. Place mouse near center of maze, facing an open arm.
2. Leave for 5 min (*see* **Note 3**).
3. Remove mouse from maze and then return to home cage.
4. Analyze data using video tracking software or "manually" from saved video recordings by an observer using a stopwatch. An arm entry can be defined as the whole body (not including the tail) of the mouse entering the area.
5. Possible data to be analyzed include time spent in each zone, number of entries to each zone, and distance moved. These parameters can be used to draw conclusions about specific changes in anxiety as previously discussed [5, 7, 22].

3.3 Open Field

1. Place mouse at any edge of the open field. Alternate start positions between mice.
2. Leave in open field to explore for 10 min.
3. Remove mouse from open field and then return to home cage.
4. Analyze data using video tracking software or "manually" by stopwatch from saved video recordings. Define a center and outer zone.
5. Possible data to be analyzed include time spent in each zone, number of entries to each zone, and distance moved. These parameters can be used to draw conclusions about specific changes in anxiety and exploratory activity as previously discussed [6, 7, 22].

3.4 Novel Object Recognition Test

1. Perform two habituation sessions, 24 h apart, as per method for Open Field.
2. 24 h following the second habituation session, the training session is conducted. Place two identical objects toward the corners of one wall of the open field, 10–15 cm from each wall (*see* Fig. 3 and **Note 4**).
3. Place mouse in open field along opposite wall to where objects are placed.
4. Leave in open field to explore for 10 min.
5. Remove mouse from open field and then return to home cage.
6. One hour later (*see* **Note 5**), the retention session is conducted. Place one object from the training session and one novel object in the same position as previously used.
7. Place mouse in open field along opposite wall to where objects are placed.
8. Leave in open field to explore for 5 min.
9. Remove mouse from open field and return it to its home cage.
10. Analyze data using video tracking software or "manually" by stopwatch from saved video recordings. Define a sniffing or interaction zone around each object as 1.5 × the size of the object.
11. Possible data to be analyzed include time the nose of mouse spent in each zone (interaction time), number of entries to each zone, and distance moved.

3.5 Y-Maze Short-Term Spatial Recognition Memory

1. Place sawdust litter in the base of the maze along with cage bedding from mice to be tested in each group. Mix between trials to reduce localized olfactory cues.
2. Training session: Block either one of the test arms of the Y-maze with a barrier (*see* Fig. 4 and **Note 5**).

3. Place mouse in start arm.
4. Leave in maze to explore for 10 min.
5. Remove mouse from maze and return it to its home cage.
6. One hour later (*see* **Note 5**), the retention session is conducted. Remove barrier so that all three arms are open.
7. Place mouse in start arm.
8. Leave in maze to explore for 5 min.
9. Remove mouse from maze and return it to its home cage.
10. Analyze data using video tracking software or "manually" by stopwatch from saved video recordings. An arm entry is defined when two thirds or the whole body (not including the tail) of the mouse is in the area.
11. Possible data to be analyzed include time mouse spent in each zone, number of entries to each zone, and distance moved.

3.6 Social Interaction Test

1. Place "stranger" enclosures in each of the two outer chambers in line with each other toward the back wall of the chamber (Fig. 5). Ensure removable barriers between chambers are closed.
2. Habituation session: Place test mouse in center chamber.
3. Open barriers and leave the mouse to explore for 5 min.
4. Sociability session: Immediately after the end of the 5 min, place the test mouse back in the center compartment with barriers closed.
5. Place stranger mouse 1 (*see* **Note 6**) into one of the stranger enclosures (*see* **Note 4**). The remaining stranger enclosure is to remain empty.
6. Open barriers and leave the test mouse to explore for 5 min.
7. Preference for social novelty session: Immediately after the end of the 5 min, place the test mouse back in the center compartment with barriers closed.
8. Place stranger mouse 2 (*see* **Note 6**) into the previously empty stranger enclosure.
9. Open barriers and leave the test mouse to explore for 5 min.
10. Remove all mice from the chamber and return them to their own home cage.
11. Analyze data using video tracking software or "manually" by stopwatch from saved videos. Define a sniffing or interaction zone around each stranger mouse as 1.5 × the size of their enclosure.
12. Possible data to be analyzed include time nose of the test mouse spent in each zone (interaction time), number of entries to each zone, and distance moved.

3.7 Forced Swim Test

1. Fill cylinders with 1.7 L of water at 23 ± 2 °C.
2. Place mice in water and leave for 6 min (*see* **Note 7**).
3. Remove from water and dry excess water off with towel.
4. Place into clean cage on heat mat until dry (*see* **Note 2**).
5. Return to home cage.
6. Analyze data using behavioral scoring software or stopwatch.
7. Time spent swimming, climbing, and immobile (*see* **Note 8**) should be recorded.

3.8 Fear Conditioning Test

1. Set up chambers; clean and apply relevant scent and any additional sawdust bedding or wall inserts required (*see* **Notes 1 and 9**).
2. Place mice individually in the chambers, and leave for a set amount of time according to previously set protocols in Video-Freeze software. Example fear protocols: Conditioning sessions include the delivery of a small number (usually 3–5) of 30 s, 5 kHz, 70–80 dB tones (conditioned stimulus (CS)) that coterminate with an unconditioned stimulus (US, mild foot shock of 0.7 mA) that lasts for 1 s.

   Two common variations of testing protocols are suggested below, although these can be modified based on research questions and strain/sex of the mice used.

   2.1. Context and cued memory: Day 1—During a 6 min conditioning phase of the protocol, following a 2.5 min habituation period, chambers automatically deliver 3 × CS paired with US with an inter-trial interval (ITI) of 30 s. Day 2—The animal is placed back in the conditioning chamber, but no tones or shock are delivered, to measure context fear memory. Day 3—The animal is placed into the alternate context, while three tones are delivered with no shock to measure cued fear memory or tone memory [25, 26].

   2.2. Extinction learning and memory: Day 1—During an 11 min conditioning phase of the protocol, following a 3 min habituation period, chambers automatically deliver 3 × CS paired with US with an ITI of 3 min. Day 2—The animal is placed into the alternate context, and 40 tones are delivered but no shock, with an ITI of 5 s to measure cued fear memory (during early tone presentations) and extinction learning (across the full session). Day 3—The same conditions as used on Day 2 are repeated to measure extinction memory consolidation [23, 27].
3. At the end of the session, remove mouse from the chamber and return it to its home cage.

4. Repeat process on subsequent days according to protocol of choice (*see* **Note 9**).
5. Analyze data using VideoFreeze software.
6. Time spent freezing during each component of the protocol should be recorded.

3.9 Locomotor (Hyper)Activity

1. Set up chambers and start software (*see* **Note 1**).
2. Place mice in chambers individually and fit lids to top of chamber.
3. Leave for set amount of time according to chosen protocol. Example LMA protocols:
   3.1. Baseline only: To determine any effects of genotype, previous or chronic treatments, or other environmental interventions on exploratory locomotor activity, place mice in chambers for 10–60 min and leave undisturbed.
   3.2. Drug-induced locomotor hyperactivity: Place mice in chambers for a habituation period of 15–60 min (this will depend on how long the strain of mice takes to decrease initial high levels of exploratory activity). Mice are then briefly removed for drug administration and returned to chamber for 1–2 h [28, 29]. The time monitored following drug administration will depend on the research question, pharmacokinetics of the drug, and time of action.
4. At end of the session, remove mice from chambers and return them to their home cage.
5. Analyze data using activity monitor software.
6. Distance travelled in 5 min time bins should be recorded if time course analysis is used.

4 Notes

1. When using video tracking software, always monitor tracking or other data being collected during behavioral testing, and make a note if it looks like there were any problems, for example, possibly aberrant behavior or the tracking software briefly “losing” the mouse. During later analysis, data from these sessions must be carefully checked for being outliers. Occasionally, checking connections and rearranging cables prior to testing can help fix problems with incorrect data collection. Equipment with tone and movement detection settings such as used for fear conditioning should be calibrated regularly to ensure correct parameters are used. Make sure to record and save all videos to allow for re-analysis offline at a later time if necessary.

2. Provided all steps are followed correctly, you should not expect any negative health outcomes following behavioral testing. However, some excess porphyrin around the eyes and nose may appear possibly due to mild stress.
3. Mice may fall off the maze when turning around on the open arms. If this happens, pick them up as soon as possible and place back on the same place on the open arm from where they fell. Make a note of this in the lab book. To assist retrieval of mice in these cases, it can be useful to place empty cages underneath the open arms or a barrier around the base of the maze. Upon analysis, the data from these mice should be carefully screened for "outlier" scores. To reduce the chance of the mouse falling off the open arms, a wall of approximately 0.5 cm high may be added on these arms.
4. It is recommended to give some mice different training and retention objects than other mice to prevent data being influenced by any object preferences. Also, alternate the side where the novel object is placed to account for any possible side preferences seen in mice. Similarly, alternate the novel arm in the Y-maze and stranger sides in the sociability test between test mice.
5. The time between training and retention sessions for NORT and Y-maze can be adjusted depending on the research question (short-term or long-term memory) and sex and "performance" of the mice based on previous experiments. On testing days, multiple successive mice can undergo training sessions until the time is ready for the first mouse to complete its retention session. Retention sessions for all mice in this group should then be completed before the next group of mice start their training sessions.
6. Stranger mice should be the same sex and background strain as the test mouse. We have used either mice of the same age or adolescent mice [19, 20]. They should not be siblings or cage mates of the test mouse, and mice used for Stranger 1 and Stranger 2 should also not be cage mates.
7. Mice in FST must be carefully monitored at all times during the experiment and removed from the beaker if the head goes underwater for more than a few seconds.
8. Immobility behavior can be characterized by the absence of any active swimming or paddling movements, often consisting of floating or freezing behaviors. Climbing behavior can be defined as upward movement of the forearms against the side of the swim cylinder [30].
9. It is recommended to have two types of chambers with different contexts. For example, one type can have the house light on and the other house light off; one type can be the standard

square shape with white back wall, and the other includes a rounded insert with added stripes; one type should have a peppermint essence scent applied, while the other should be wiped with water but have sawdust bedding underneath the grid floor. Animals should perform conditioning sessions and context memory testing sessions in one context but be switched to the alternate context for cued memory testing sessions. Half of the test animals undergo training in one context and the remaining animals in the alternate context to prevent any effects of context on baseline freezing.

References

1. Petković A, Chaudhury D (2022) Encore: Behavioural animal models of stress, depression and mood disorders. Front Behav Neurosci 16: 931964
2. van den Buuse M, Garner B, Gogos A, Kusljic S (2005) Importance of animal models in schizophrenia research. Aust N Z J Psychiatry 39: 550–557
3. Chadman KK, Yang M, Crawley JN (2009) Criteria for validating mouse models of psychiatric diseases. Am J Med Genet B Neuropsychiatr Genet 150b:1–11
4. van den Buuse M (2010) Modeling the positive symptoms of schizophrenia in genetically modified mice: pharmacology and methodology aspects. Schizophr Bull 36:246–270
5. Wall PM, Messier C (2001) Methodological and conceptual issues in the use of the elevated plus-maze as a psychological measurement instrument of animal anxiety-like behavior. Neurosci Biobehav Rev 25:275–286
6. Denenberg A (2006) Open-field behavior in rats: what does it mean? Ann N Y Acad Sci 159:852–859
7. Jaehne EJ, Kent JN, Lam N, Schonfeld L, Spiers JG, Begni V, De Rosa F, Riva MA, van den Buuse M (2023) Chronic running-wheel exercise from adolescence leads to increased anxiety and depression-like phenotypes in adulthood in rats: effects on stress markers and interaction with BDNF Val66Met genotype. Dev Psychobiol 65:e22347
8. Al Dahhan NZ, De Felice FG, Munoz DP (2019) Potentials and pitfalls of cross-translational models of cognitive impairment. Front Behav Neurosci 13:48
9. Dellu F, Contarino A, Simon H, Koob GF, Gold LH (2000) Genetic differences in response to novelty and spatial memory using a two-trial recognition task in mice. Neurobiol Learn Mem 73:31–48
10. Dellu F, Mayo W, Cherkaoui J, Le Moal M, Simon H (1992) A two-trial memory task with automated recording: study in young and aged rats. Brain Res 588:132–139
11. Dere E, Huston JP, De Souza Silva MA (2007) The pharmacology, neuroanatomy and neurogenetics of one-trial object recognition in rodents. Neurosci Biobehav Rev 31:673–704
12. Commons KG, Cholanians AB, Babb JA, Ehlinger DG (2017) The rodent forced swim test measures stress-coping strategy, not depression-like behavior. ACS Chem Neurosci 8:955–960
13. Leite-Almeida H, Castelhano-Carlos MJ, Sousa N (2021) New horizons for phenotyping behavior in rodents: the example of depressive-like behavior. Front Behav Neurosci 15: 811987
14. Kaidanovich-Beilin O, Lipina T, Vukobradovic I, Roder J, Woodgett JR (2011) Assessment of social interaction behaviors. J Visual Exp 2011:2473
15. Yang M, Silverman JL, Crawley JN (2011) Automated three-chambered social approach task for mice. Curr Protoc Neurosci Chapter 8: Unit 8.26
16. O'Tuathaigh CM, Kirby BP, Moran PM, Waddington JL (2010) Mutant mouse models: genotype-phenotype relationships to negative symptoms in schizophrenia. Schizophr Bull 36: 271–288
17. Millan MJ, Bales KL (2013) Towards improved animal models for evaluating social cognition and its disruption in schizophrenia: the CNTRICS initiative. Neurosci Biobehav Rev 37:2166–2180
18. Corrone M, Ratnayake R, De Oliveira N, Jaehne EJ, van den Buuse M (2023) Methamphetamine-induced locomotor sensitization in mice is not associated with deficits in a range of cognitive, affective and social

behaviours: interaction with brain-derived neurotrophic factor (BDNF) Val66Met. Behav Pharmacol 34:20–36

19. Genders SG, Scheller KJ, Jaehne EJ, Turner BJ, Lawrence AJ, Brunner SM, Kofler B, van den Buuse M, Djouma E (2019) GAL_3 receptor knockout mice exhibit an alcohol-preferring phenotype. Addict Biol 24:886–897
20. Jaehne EJ, Ameti D, Paiva T, van den Buuse M (2017) Investigating the role of serotonin in methamphetamine psychosis: unaltered behavioral effects of chronic methamphetamine in 5-HT_{1A} knockout mice. Front Psych 8:61
21. Tran SC, Jaehne EJ, Dye LE, Wong J, Bakas JS, Gasperoni JG, Hale MW, van den Buuse M, Dworkin S, Grommen SVH, De Groef B (2021) Effect of pleomorphic adenoma gene 1 deficiency on selected behaviours in adult mice. Neuroscience 455:30–38
22. Lad HV, Liu L, Paya-Cano JL, Parsons MJ, Kember R, Fernandes C, Schalkwyk LC (2010) Behavioural battery testing: evaluation and behavioural outcomes in 8 inbred mouse strains. Physiol Behav 99:301–316
23. Jaehne EJ, Kent JN, Antolasic EJ, Wright BJ, Spiers JG, Creutzberg KC, De Rosa F, Riva MA, Sortwell CE, Collier TJ, van den Buuse M (2022) Behavioral phenotyping of a rat model of the BDNF Val66Met polymorphism reveals selective impairment of fear memory. Transl Psychiatry 12:93
24. Kokras N, Baltas D, Theocharis F, Dalla C (2017) Kinoscope: an open-source computer program for behavioral pharmacologists. Front Behav Neurosci 11:88
25. Chen ZY, Jing D, Bath KG, Ieraci A, Khan T, Siao CJ, Herrera DG, Toth M, Yang C, McEwen BS, Hempstead BL, Lee FS (2006) Genetic variant BDNF (Val66Met) polymorphism alters anxiety-related behavior. Science 314:140–143
26. Notaras M, Hill R, Gogos JA, van den Buuse M (2016) BDNF Val66Met genotype determines hippocampus-dependent behavior via sensitivity to glucocorticoid signaling. Mol Psychiatry 21:730–732
27. Gururajan A, Hill RA, van den Buuse M (2015) Brain-derived neurotrophic factor heterozygous mutant rats show selective cognitive changes and vulnerability to chronic corticosterone treatment. Neuroscience 284:297–310
28. Manning EE, Halberstadt AL, van den Buuse M (2016) BDNF-deficient mice show reduced psychosis-related behaviors following chronic methamphetamine. Int J Neuropsychopharmacol 19:pyv116
29. van den Buuse M, Ruimschotel E, Martin S, Risbrough VB, Halberstadt AL (2011) Enhanced effects of amphetamine but reduced effects of the hallucinogen, 5-MeO-DMT, on locomotor activity in 5-HT_{1A} receptor knockout mice: implications for schizophrenia. Neuropharmacology 61:209–216
30. Slattery DA, Cryan JF (2012) Using the rat forced swim test to assess antidepressant-like activity in rodents. Nat Prot 7:1009–1014

Chapter 8

A Simple Method for Quantifying Larval Locomotion in *Drosophila melanogaster*

Jiayi Lin, Sarah Mele, Matthew D. W. Piper, and Travis K. Johnson

Abstract

The fruit fly *Drosophila melanogaster* is a powerful genetic model that has been used for many decades to study nervous system function, development, and behavior. There are a large number of developmental and behavioral traits that can be measured to provide a broad readout of neurological function. These include patterned motor behaviors, such as larval locomotion, which can be used to assess whether genetic or environmental factors affect nervous system function to provide an entry point for deeper mechanistic studies. Here, we describe a protocol for quantifying larval locomotion using a simple camera setup and a freely available image analysis software. This protocol can be readily applied to human disease models or in toxicology studies, for example, to broadly assess the impact of treatments on neurological function.

Key words *Drosophila melanogaster*, Larval crawling assay, Locomotor activity, Larval locomotion, Behavioral assay

1 Introduction

Over the past century, research using the genetic model organism *Drosophila melanogaster* has contributed enormously to our fundamental understanding of developmental biology and neurobiology. From this work, we have learned that the cellular framework that underpins neuronal connectivity and function is highly conserved between *Drosophila* and mammals. The relative ease with which *Drosophila* neurobiology can be investigated, and the sophisticated genetic tools at its disposal, has led to a range of foundational discoveries regarding the molecular mechanisms and general principles of neurogenesis, axon guidance, neurotransmission, circadian rhythm, learning and memory, and complex behavior [1]. With regard to complex behavior, *Drosophila* has been particularly valuable for studying neural circuits that coordinate motor behaviors such as walking and crawling. This includes studies of synaptic

Sebastian Dworkin (ed.), *Neurobiology: Methods and Protocols*, Methods in Molecular Biology, vol. 2746,
https://doi.org/10.1007/978-1-0716-3585-8_8,

transmission, particularly at the larval neuromuscular junction, as well as a focus on central pattern generation for rhythmic outputs and the integration of sensory inputs [2].

As many human diseases manifest with neurological symptoms, *Drosophila* can be an excellent model to assess the impact of disease pathogenesis on the nervous system [3–6]. Larval locomotor behavior has been widely investigated in this context as it offers a sensitive measure of nervous system impairment [7]. It is particularly useful when disease onset occurs during the larval stage and individual arrest prior to adulthood.

The protocol we describe is adapted from a previous method [8]. In brief, larvae are introduced to a Petri dish containing colored medium for visual contrast. Larval movement is recorded using a simple camera setup and subsequently analyzed by the wrMTrck plugin in ImageJ, a freely available image analysis software. This protocol can be used to assess both second and third-instar stage larvae.

2 Materials

2.1 Recommended Equipment

1. Grade 1 round pony/goat hair paintbrush.
2. 100-micron mesh cloth filter.
3. 200 mL beaker
4. Petri dishes (60 mm for second-instar, 90 mm for third-instar larvae).
5. Kimwipes.
6. Glass stirring rod.
7. Disposable Pasteur pipettes.
8. Paper towel.
9. Masking tape.
10. Webcam or smartphone camera.
11. Dark, nonreflective surface, e.g., black cloth.

2.2 Solutions and Chemical Components

1. Sucrose: 20% w/vol in Milli-Q water.
2. Phosphate-buffered saline (PBS; 10×) pH 7.4: 80 g NaCl, 2 g KCl, 14.4 g $NaHPO_4$, 2.4 g KH_2PO_4, distilled water to 1 L. Adjust pH to 7.4 using 37% HCl or 10 M NaOH. For 1× working solution, take 100 mL from 10× PBS and dilute to 1 L with distilled water.
3. Apple juice agar: Add 2024 mL apple juice (*see* **Note 1**) to a 5 L jug on a stirrer. Add 24 mL double-distilled water to the jug. Add 130.3 g sucrose and 261 g dextrose, and allow to dissolve. Add 15 mL NaOH. Dispense solution to 250 mL Schott

bottles (approximately 200 mL each). To each bottle, add 8 g agar and autoclave (*see* **Note 2**). Apple juice agar can be stored at room temperature after autoclaving.

4. Blue food dye, e.g., Queen blue food color (1.8% dyestuff) (*see* **Note 3**).
5. Propionic acid.

3 Methods

3.1 Preparation of Food Plates

1. Microwave 200 mL apple juice agar, being careful not to boil over.
2. Add 500 μL propionic acid, and stir thoroughly with a glass stirring rod.
3. Add 1 mL blue food dye to 200 mL apple juice agar.
4. Stir thoroughly with a glass stirring rod.
5. Pour into Petri dishes. Approximately 25 plates (60 mm) or 10 plates (90 mm) can be made from 200 mL apple juice agar (see **Note 4**).

3.2 Larval Collection and Video Acquisition

1. To isolate larvae for the crawling assay, pour 20% sucrose solution into a vial containing larvae at the desired developmental stage until the vial is about half-full (*see* **Note 5**).
2. Gently disrupt the top layer of the medium with the paintbrush to release the larvae from the food, allowing them to float to the surface of the sucrose.
3. Isolate the floating larvae by pouring the sucrose solution through the 100-micron cloth mesh or sieve into the beaker.
4. Wash sucrose and food residue from the larvae thoroughly with 1× PBS using a disposable Pasteur pipette. Gently remove any residue food from the mesh using a paintbrush. Place the mesh with larvae on top on a paper towel and carefully blot the mesh dry from underneath.
5. Place the Petri dish on a dark, nonreflective surface on a lab bench with consistent lighting (*see* **Note 6**).
6. Gently transfer 5 larvae from the fine cloth mesh/sieve to the center of a Petri dish containing colored apple juice agar using a paintbrush (*see* **Notes** 7 and **8**).
7. Allow larvae to acclimate to the new environment for 3 min before filming (*see* **Note 9**).
8. Place the webcam or smartphone camera at a sufficient height so that Petri dish fills the entire field of view (*see* **Note 10)** (Fig. 1).
9. Record for 3 min (*see* **Note 11**).

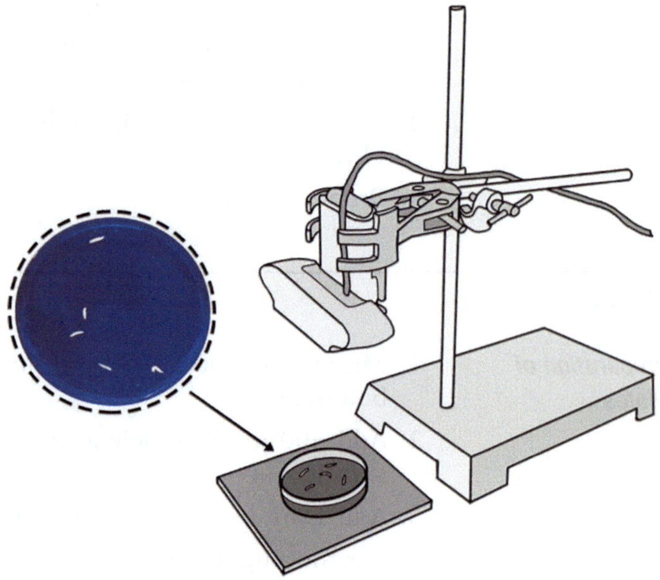

Fig. 1 Camera setup. Position the agar plate on a dark surface, in a location where the surface of the agar is evenly lit. Lower the camera such that the agar plate fills the field of view. A phone camera can be used in place of a webcam. Third-instar larvae are pictured on a 60 mm agar plate

3.3 Video Processing and Data Collection

1. Upload the video files to a computer and convert files to .avi format using FFmpeg (*see* **Note 12**).
2. Open the .avi file in ImageJ, enter 5400 for "last frame" to generate a set of 5400 frames (substacks), and check "Convert to grayscale" (Fig. 2a). This will crop the clip to a 3-min video.
3. Use the oval selection tool in the toolbar to outline the Petri dish field and crop the video (Image -> Crop) (*see* **Note 13**) (Fig. 2b).
4. Use the straight line tool to draw a line from one edge of the Petri dish to the other, and then, measure and record the diameter of the Petri dish in pixels (Analyze -> Measure). This measurement is required for calculating a conversion factor in **step 11**.
5. Deselect objects (Edit -> Selection -> Select none) and invert the stack of images (Edit -> Invert) to show the larvae as dark objects on a light background (Fig. 2c).
6. Remove the background from all images (Process -> Subtract background; Rolling ball radius 1; check "Light background" and "Sliding paraboloid") (Fig. 2d).
7. Set the threshold (Image -> Adjust -> Threshold). Select "MaxEntropy" mode and "B&W" and adjust the parameters by dragging the parameter sliders until the larvae are white on a

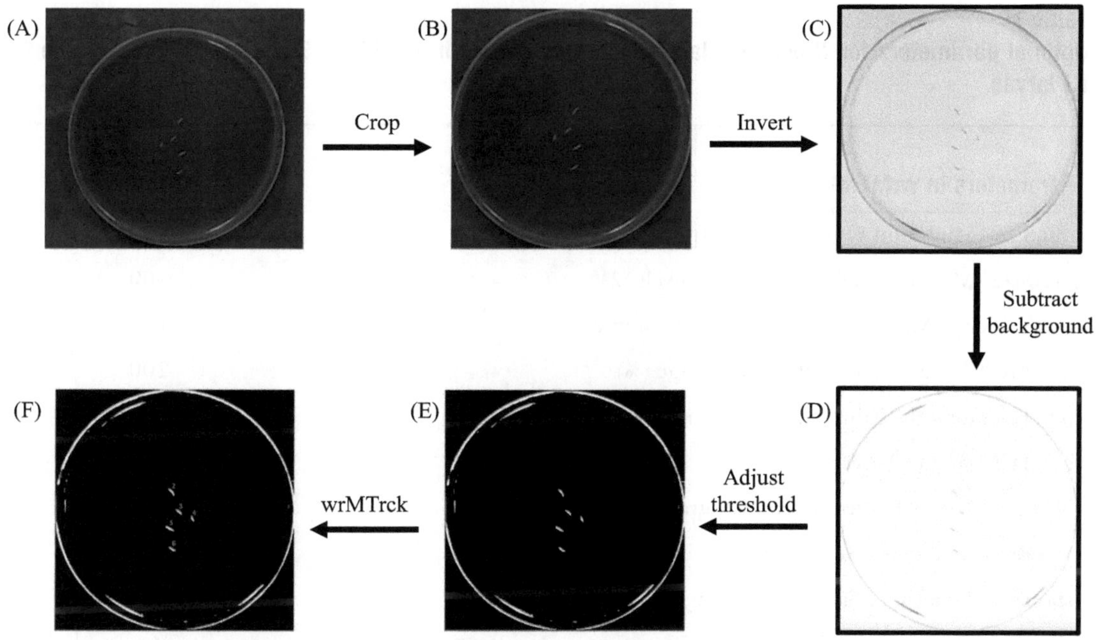

Fig. 2 Video processing in ImageJ. (**a**) The avi. video is opened in imageJ and converted to grayscale. (**b**) Crop the video using the oval selection tool to minimize the file size and (**c**) invert the stack of images so the larvae are shown as dark on a light background. (**d**) Remove the background from all images and (**e**) adjust the threshold so the larvae appear white on a black background. (**f**) After running the wrMTrck plugin, a video with numbers assigned to each larva is generated

black background. Select "Apply" and then deselect "Calculate threshold for each image" (Fig. 2e).

8. Run the wrMTrck plugin (Plugin -> wrMTrck) with the parameters listed in Table 1. These parameters work well for third-instar larvae. For second-instar larvae, parameters may require adjustment (*see* **Note 14**).
9. A new video with numbers assigned to single larvae is generated (Fig. 2f), along with a results table and a figure that shows the paths of all larvae. Export the results table (*see* **Notes 15–18**).
10. To convert the measurements recorded in the results table (Length, Distance, AveSpeed, MaxSpeed) into millimeters (mm) and millimeters/second (mm/sec), a conversion factor (mm/pixel) is calculated. The diameter of the Petri dish in millimeters is divided by the diameter in pixels (measured in **step 5**). For "Length" and "AvgSpeed" data in the results table, multiply the length or average speed in pixels by the conversion factor to calculate the total length in millimeters or average speed in mm/sec travelled.

Table 1
Optimal parameters for *Drosophila* larval locomotor assay in wrMTrck. Size settings appropriate for L3 larvae

Parameters in wrMTrck	Optimal settings
minSize--Minimum Objective Area (pixels^2)	10
maxSize--Maximum Objective Area (pixels^2)	400
maxVelocity--Maximum Velocity (pixels/frame)	10
maxAreaChange--Maximum area change (%)	200
minTrackLength--Minimum track length (frames)	500
bendThreshold--Threshold for turn	2.0
binSize - Size of bin for speed histogram (pixels/frame) (0 = disable)	0.0
saveResultFile--Save Result File	--
showPathLengths--Display Path Lengths	✓
showLabels--Show Labels	✓
showPositions--Show Positions	--
showPaths--Show Paths	✓
showSummary--show a summary of tracking	✓
roundCoord--round off coordinates	--
Smoothing--point smoothing	✓
plotBendTrack--Quality control plots for thrashing analysis	--
rawData--(0 = off, 1 = XYcord, 2 = Ellipse, 3 = AreaPerimDist, 4 = Ellipse+Circ, 5 = Bendcalc)	0
bendDetect--(0 = off, 1 = Angle, 2 = AspectRatio, 3 = AR + Histogram)	2
FPS--frames/sec (0 = try to load from file)	0
backSub--On-the-fly background sutration (0 = off, 1 = F1RB15)	0
threshMode--Thresholding method (only if backSub>0)	Otsu
fontSize--Size of labelling font	16

To change this: (*See* **Note 14**)

4 Notes

1. We use reconstituted apple juice (99.7%) which contains citric acid and vitamin C.
2. Gelling properties of agar can vary between suppliers, and ensure the agar is sufficiently firm such that larvae cannot burrow easily.
3. Other dark color food dye (e.g., black) can also be used.

4. Food plates can be prepared ahead of time and refrigerated. When using the food plates, ensure they are at room temperature and dry (use a Kimwipe to absorb moisture).
5. Behavior can be very sensitive to environmental conditions. Ensure larval cultures are kept under uniform density, temperature, and photoperiod for consistency.
6. Avoid bright reflections from appearing on the surface of the agar by positioning the plate away from direct lightning. Bright areas can be mistakenly labelled as larvae by wrMTrck.
7. Initially, we recommend analyzing five replicate plates each with five larvae per plate for a given condition. In our experience, this is usually sufficient to detect consistent locomotor impairments as well as to reduce errors in the wrMTrck analysis.
8. Do not reuse agar plates as the larvae leave a trail of residue that may influence the behavior of other larvae. Transferring larvae also disturbs the surface of the agar over time.
9. Larvae tend to be more active when they are first introduced to the Petri dish. Three minutes is usually sufficient for larvae to acclimate to the medium and achieve stable locomotion.
10. The webcam can be fixed to a retort stand at an appropriate height using masking tape. If a smartphone is used, position it on a stand or the edge of a box raised to an appropriate height.
11. Record video at no higher than 30 frames per second to avoid subsequent problems in data processing. Recording is limited to 3 min to minimize the occurrence of larvae burrowing at the edge of the plate where they escape capture by the camera and cannot be analyzed.
12. Note that the video format can also be converted by other freely available video format conversion software. For FFmpeg, download (https://www.ffmpeg.org/). Windows instructions:

    Create a folder in This PC > Local Disk (C:) named FFmpegTool and extract all the files in this folder. Put your video files in the directory C:\FFmpegTool\bin. Run FFmpeg by searching "Run" in windows, type "cmd," and press enter. In the command line interface, type "cd\ followed by "cd FFmpeg-Tool\bin." For each video recording, run the following commands: ffmpeg -i **FILENAME.filetype** -an **NEWFILENAME.filetype** to remove audio and then "ffmpeg -i **FILENAME.filetype** -f avi -vcodec mjpeg **NEWFILENAME.**avi" to convert to AVI NV12 format. The AVI NV12 format is required to open the video file in ImageJ.

    For MAC, run FFmpeg by opening Terminal and type "ffmpeg" in the command line interface. Then, type "cd" followed by location of the video (e.g., cd desktop). To convert video format, enter commands "ffmpeg -i **FILENAME.filetype** -pix_fmt nv12 -f avi -vcodec rawvideo **NEWFILENAME**.avi."

13. The purpose of this step is to minimize the file size.
14. To identify second-instar larvae, change the minSize and maxSize parameters to reflect their size. Use the straight line tool to measure the larval size in pixels (Analyse -> Measure).
15. Relevant parameters in the output table: length, total length of the track (speed * time); distance, distance between start point to end point of the track; time(s), time taken to complete the track; MaxSpeed, AvgSpeed.
16. When two or more larvae crawl near to one another, their labels can become switched. This can be seen in the labelled video. In this case, the simplest solution is to discard the data for these larvae. WrMTrck will record two unbroken tracks, and therefore, they cannot easily be corrected in data processing.
17. If the label of a larva changes during the video (due to light flickering), data for these labels can be combined as follows:

    Total length = Length of label 1 + Length of label 2 + . . . + Length of label n;

    Maximum speed, compare the MaxSpeed of all labelling and record the largest value;

    Average speed = (Length of label 1 + Length of label 2 + . . . + Length of label n)/(Time of label 1 + Time of label 2 + . . . + Time of label n).
18. If a larva burrows into the edge of the Petri dish and cannot be captured for more than 5 s, the data for the larva should not be recorded.

References

1. Bellen HJ, Tong C, Tsuda H (2010) 100 years of *Drosophila* research and its impact on vertebrate neuroscience: a history lesson for the future. Nat Rev Neurosci 11:514–522. https://doi.org/10.1038/nrn2839
2. Clark MQ, Zarin AA, Carreira-Rosario A, Doe CQ (2018) Neural circuits driving larval locomotion in *Drosophila*. Neural Dev 13:1–10. https://doi.org/10.1186/s13064-018-0103-z
3. Bolus H, Crocker K, Boekhoff-Falk G, Chtarbanova S (2020) Modeling neurodegenerative disorders in *Drosophila melanogaster*. Int J Mol Sci 21:3055. https://doi.org/10.3390/ijms21093055
4. Link N, Bellen HJ (2020) Using *Drosophila* to drive the diagnosis and understand the mechanisms of rare human diseases. Development 147. https://doi.org/10.1242/dev.191411
5. Ugur B, Chen K, Bellen HJ (2016) *Drosophila* tools and assays for the study of human diseases. Dis Model Mech 9:235–244. https://doi.org/10.1242/dmm.023762
6. Pandey UB, Nichols CD (2011) Human disease models in *Drosophila melanogaster* and the role of the fly in therapeutic drug discovery. Pharmacol Rev 63:411–436. https://doi.org/10.1124/pr.110.003293
7. Tsai HY, Wu SC, Li JC, Chen YM, Chan CC, Chen CH (2020) Loss of the *Drosophila* branched-chain α-ketoacid dehydrogenase complex results in neuronal dysfunction. Dis Model Mech 13. https://doi.org/10.1242/dmm.044750
8. Brooks DS, Vishal K, Kawakami J, Bouyain S, Geisbrecht ER (2016) Optimization of wrMTrck to monitor *Drosophila* larval locomotor activity. J Insect Physiol 93:11–17. https://doi.org/10.1016/j.jinsphys.2016.07.007

Chapter 9

Neural Stem/Progenitor Cell (NSPC) Extraction and Culture

Jemma Gasperoni and Sebastian Dworkin

Abstract

Neural stem–progenitor cells (NSPCs) are multipotent, self-renewing cells that generate radial glial cells (RGC). RGCs then give rise to neurons and glia during neural development. Here, we describe the process of NSPC isolation and culturing to form clonal aggregates termed neurospheres. There are multiple assays outlined in this chapter that allow us to quantify differences in proliferation, self-renewal potential, and differentiation of these cells.

Key words Stem cells, Brain, Mouse, Neurons, Neurosphere, Cell culture, Differentiation

1 Introduction

In the field of stem and progenitor cell research, understanding the intricate processes that govern neurobiology is essential in order to understand functional pathways within the brain. Stem and progenitor cells possess the remarkable ability to differentiate into various cell types within the nervous system, making them invaluable for studying neurodevelopment and potential therapeutic applications for neurodegenerative disease or trauma-related injury. However, comprehending the underlying mechanisms that govern their behavior necessitates reliable and versatile experimental techniques.

One such technique that has gained significant recognition in recent years is the neurosphere assay. Originally developed in 1992 [1], this assay has since become a cornerstone in neural stem cell research. It enables the isolation, expansion, and characterization of neural stem cells from diverse regions of the central nervous system, including the embryonic, fetal, and adult stages [2–4]. Moreover, the assay allows for the examination of stem and progenitor cell responses to various genetic and pharmacological manipulations, offering valuable insights into neurodevelopmental processes, neuroregeneration, and disease modelling [5–7]. These neurospheres

Sebastian Dworkin (ed.), *Neurobiology: Methods and Protocols*, Methods in Molecular Biology, vol. 2746, https://doi.org/10.1007/978-1-0716-3585-8_9,

are composed of a heterogeneous population of cells that encompass neural stem and progenitor cells that possess the capacity to differentiate into the main cell types of the central nervous system, including neurons, astrocytes, and oligodendrocytes. By utilizing specific growth factors and culturing conditions, researchers can effectively maintain the self-renewal capacity of stem and progenitor cells, promoting their proliferation and preventing spontaneous differentiation.

This assay provides numerous advantages for studying stem and progenitor cell neurobiology. Firstly, the neurosphere assay allows for the expansion of neural stem and progenitor cells in large quantities while preserving their multipotency. Secondly, the neurosphere assay offers a three-dimensional (3D) culture system that more closely resembles the in vivo microenvironment. The 3D nature of neurospheres provides a supportive "in vitro niche" for stem and progenitor cells, mimicking the cellular interactions and spatial organization present in the nervous system [2]. Thirdly, the neurosphere assay allows for the examination of stem and progenitor cell responses to a variety of factors and stimuli. By modulating the composition of the culture medium, researchers can investigate the effects of growth factors, small molecules, and other biological cues on stem cell behavior [8, 9]. Additionally, the assay enables the introduction of genetic manipulations, such as gene overexpression, deletion, or knockdown, to dissect the molecular pathways involved in stem and progenitor cell neurobiology [2]. The neurosphere assay also offers an opportunity for drug screening and toxicity testing, contributing to the development of novel treatment strategies. Lastly, the ability of these NSPCs to differentiate can also be modelled in vitro, thereby determining if a particular progenitor population may be compromised in extracted cells.

In summary, the neurosphere assay has emerged as an invaluable tool for understanding stem and progenitor cell neurobiology. Its ability to expand and maintain multipotent cells, its 3D culture environment, and its versatility in studying various factors and diseases make it an ideal system for unravelling the complex mechanisms governing neural development, regeneration, and disease progression.

2 Materials

2.1 NSPC Isolation and Culture

1. Phosphate-buffered saline (PBS): For a 10× solution, add 80 g NaCl, 2 g KCl, 14.4 g Na_2HPO_4 to 800 mL of distilled water. Adjust to pH 7.4 and fill to 1 L. For 1× PBS, add 100 mL of 10× PBS to 900 mL of distilled water.
2. Hanks' Balanced Salt Solution (HBSS).

3. Neurosphere basal medium (NSBM; 1:1 Dulbecco's Modification of Eagle's Medium [DMEM] and F12, 4 μg/mL heparin, 100 μg/mL penicillin/streptomycin).
4. Basic fibroblast growth factor (bFGF).
5. Epidermal growth factor (EGF).
6. B27 (without vitamin A).
7. Low-adhesion 6-well tissue culture plates.
8. Neurosphere dissociation solution (50 mL HBSS, 0.01 g EDTA, 0.0125 g trypsin [0.25 mg/mL], 0.119 g N-2-hydroxyethylpiperazine-N′-2-ethanesulfonic acid [HEPES], pH 8; frozen at −20 °C as 1 mL aliquots and thawed immediately prior to use).
9. Neutralization solution (50 mL HEPES-buffered DMEM, 0.007 g soybean trypsin inhibitor [0.14 mg/mL], frozen at −20 °C as 1 mL aliquots and thawed immediately prior to use).
10. 60-well Terasaki cell culture plates.
11. Dimethyl sulfoxide (DMSO).

2.2 Neurosphere Differentiation

1. 8-well chamber slides.
2. Poly-D-lysine and laminin.
3. Fetal calf serum (FCS).
4. 4% paraformaldehyde (PFA): Mix 4 g of PFA in approximately 70 mL deionized sterile water. Dissolve at 60 °C with shaking. Once dissolved, filter PFA, and top up to final volume of 100 mL with sterile deionized water.
5. Primary antibodies to identify neurons (mouse anti-Tuj1) and astrocytes (rabbit anti-GFAP).
6. Secondary antibodies Goat anti-Mouse 555 and Goat anti-Rabbit 488.
7. 4′,6-diamidino-2-phenylindole (DAPI).
8. Aqueous mounting medium.

2.3 Retroviral Transduction

1. 0.5 M $CaCl_2$.
2. Human embryonic kidney (HEK) cells stably transduced with retroviral gag and pol genes (HEK-293gp; viral packaging cell line).
3. DMEM supplemented with 5% FCS and 100 μg/mL penicillin/streptomycin (DMEM/FCS/PS).
4. $60 cm^2$ tissue culture plates.
5. DMEM supplemented with 5–10% FCS and 100 μg/mL penicillin/streptomycin.
6. 2× HEPES buffered-saline, pH 7.0 (HeBS).
7. Retroviral vector comprising protein of interest (e.g., MSCV).

8. Vesicular stomatitis virus glycoprotein (VSVG) envelope plasmid.
9. 5 mL syringe and 0.45 μm syringe filter.
10. Ultracentrifuge and 10 mL ultracentrifuge tube.
11. Recombinant fibronectin (rFN) in H_2O at a concentration of 1 mg/mL.
12. PBS supplemented with 2% bovine serum albumin (BSA).
13. 24-well tissue culture plates.

3 Methods

3.1 Embryonic NSPC Isolation

1. Euthanize pregnant dams at E14.5, collect embryos from the womb, and immediately place in a Petri dish containing cold PBS.
2. One at a time, de-chorionate the embryos and remove placentas (*see* **Note 1**).
3. Dissect the head of the embryo and gently cut away the skull from the brain under a dissecting microscope.
4. Next, cut the brain sagittally to separate into left and right hemispheres.
5. Remove the external cortex from each hemisphere, leaving the medial, caudal, and lateral ganglionic eminences (*see* **Note 2**).
6. Once dissected, place the ganglionic eminences into a 15 mL Falcon tube containing 5 mL of fresh HBSS on ice, until all brains have been dissected (Fig. 1).
7. In a laminar flow hood, disaggregate the tissue by gentle trituration using a pipette.
8. Centrifuge the disaggregated tissue for 5 min at 1200 × *g*.
9. Remove the supernatant and resuspend the cells in 5 mL NSBM, supplemented with 10 ng/mL basic fibroblast growth factor, 20 ng/mL epidermal growth factor, and 1:50 B27 without vitamin A.
10. Transfer the cells to a low-adhesion 6-well plate to allow the cells to form clonal aggregates (neurospheres).
11. Incubate cells at 37 °C and 5% CO_2.

3.2 NSPC Culture

1. Every 2 days, remove 2.5 mL (half volume) of media (taking care to avoid discarding neurospheres; *see* **Note 3**).
2. Add 2.5 mL of fresh NSBM, containing 1:50 B27 without vitamin A and twice the concentration of EGF and bFGF (20 ng/mL βFGF, 40 ng mL EGF; termed NSBM +2 × GFs). This will ensure that the final concentration of GFs remains at 1×.

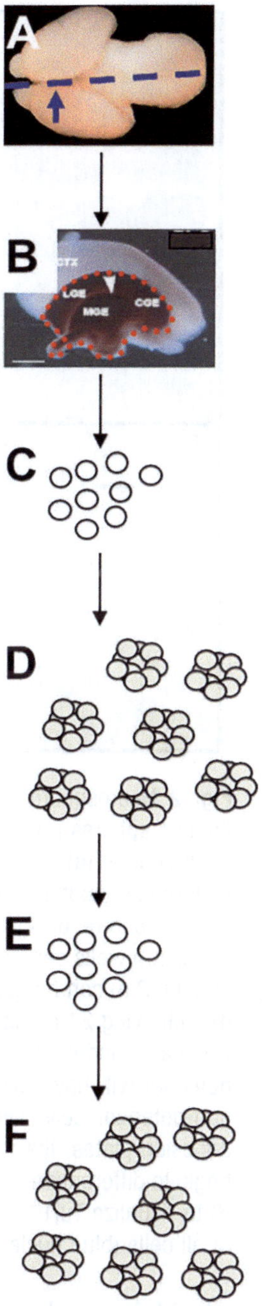

Fig. 1 Extraction of NSPCs from embryonic day 14.5 (E14.5) mouse brains. Following embryo harvest, the brains are removed from the skull (**a**). Next, the cortical layers are removed, leaving only the lateral, medial, and caudal ganglionic eminences (LGE, MGE, and CGE, respectively; **b**). This is then broken up gently with a pipette, leaving a single cell suspension (**c**). Following culture, the neural stem/progenitor cells (NSPCs) present in the GEs form clonal aggregates termed neurospheres (**d**). These can be dissociated ("passaged") into further single cell suspensions (**e**) and replated, where they will again reform neurospheres (**f**)

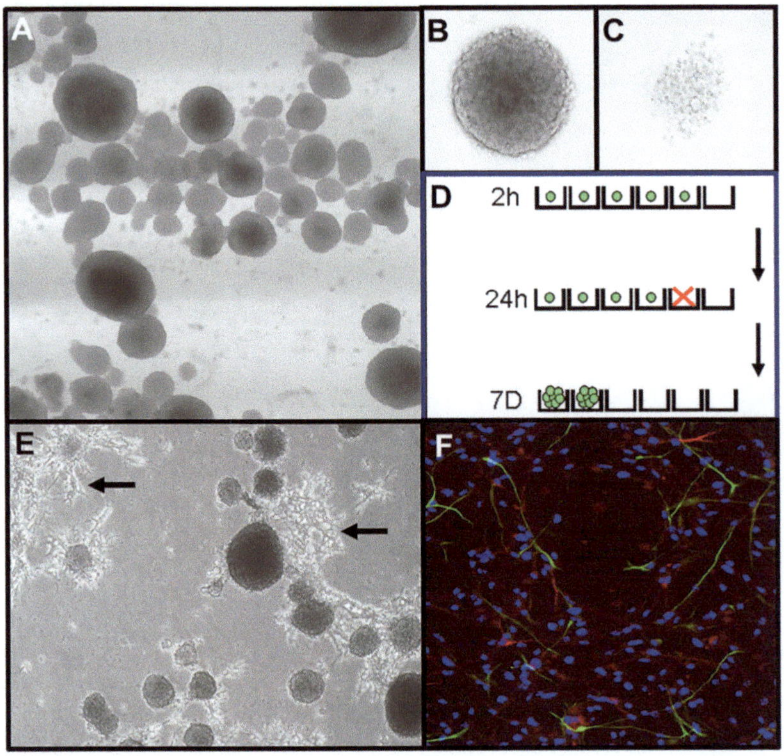

Fig. 2 Neurosphere culture techniques. Normal neurospheres grow as free-floating spheres (**a**); healthy, live neurospheres will appear as a solid ball with a dark center (**b**), whereas neurospheres in which the vast majority of cells have undergone apoptosis or necrosis will appear loose and pale (**c**). Healthy neurospheres can be dissociated and plated at a density of 5 cells/10 μL in Terasaki plates (60 wells per plate). The number of viable cells present immediately after plating (2 h) can be quantitated—most wells will have 0 or 1 cells (green dots; **d**). Following 24 h culture, few cells will have started to divide, although some will have undergone apoptosis (red x). By 7D, any cells with self-renewal potential will have given rise to neurospheres comprising >4 cells, whereas non-potential cells will have undergone apoptosis. When grown on normal adhesion plates, the neurospheres will adhere, sprout processes (arrows), and begin to differentiate (**e**). This can then be detected using immunohistochemistry (**f**) to visualize Tuj1$^+$ neurons (red) and GFAP$^+$ astrocytes (green) as a proportion of all cells (blue nuclei; DAPI)

3.3 Neurosphere Dissociation and Passaging

1. Maintain the culture until the majority of neurospheres appear large, with dark centers (approximately 5–7 days growth; Fig. 2). The media will begin to turn orange/yellow.
2. To passage, transfer the cells into a 15 mL Falcon tube and centrifuge for 7 min at 700 × *g*.
3. Remove the supernatant and gently resuspend neurospheres in 500 μL of dissociation solution for 3 min at RT with gentle trituration.

4. Neutralize the dissociation solution with 500 μL of neutralization solution for 3 min at RT with gentle trituration.
5. Next, wash dissociated cells by adding 8 mL of NSBM and centrifuge for 5 min at 1300 × *g*.
6. Remove the supernatant and resuspend cells in 1 mL of NSBM for cell counts.
7. Take the desired number of cells (depending on the assay being performed; see below) and add to 5 mL of NSBM + 1× GF. For routine passage, cells should be plated at at a density of 1–5 × 10^5 cells/ml.
8. Plate the newly passaged cells into a fresh well of a 6-well low-adhesion tissue culture plate and incubate at 37 °C and 5% CO_2.

3.4 Cumulative Cell Number Assay

1. Following neurosphere dissociation as per Subheading 3.2, **step 2**, count cells and plate at a concentration of 1–5 × 10^5 cells/mL on low-adhesion 6-well plates in NSBM + 1 × GFs.
2. Every 2–3 days, remove the media and replace with fresh NSBM + 2 × GFs. This will ensure cells continue to be grown in NSBM + 1 × GFs.
3. Every 7 days, passage cells as per Subheading 3.2, **step 2** and record weekly count. This should be repeated for 5–6 weeks depending on health of neurospheres.

3.5 Neurosphere Survival Assay

1. Plate ~20–50 day 7 neurospheres per well on low-adhesion 6-well plates in NSBM +1 × GFs.
2. Count the exact number of neurospheres present.
3. Leave neurospheres for 14 days without changing media or re-adding growth factors.
4. Count the exact number of neurospheres that are comprised of still-viable cells (Fig. 2).

3.6 Clonal Density Assay to Measure Neurosphere Formation Independent of "In Vitro Niche"

1. Dissociate neurospheres as per Subheading 3.2, **step 2**, and perform cell counts on dissociated cells.
2. Plate 1–2 × 10^4 cells per well on low-adhesion 6-well plates in NSBM + 1 × GFs.
3. Every 2 days, remove and replace the media with fresh NSBM + 2 × GFs. This will ensure cells continue to be grown in NSBM + 1 × GFs.
4. On day 7, count the number of neurospheres that have formed.

3.7 Single-Cell Assay to Determine Neurosphere Forming Potential

1. Dissociate neurospheres as per Subheading 3.2, **step 2**, and perform cell counts on dissociated cells.
2. Resuspend cells at a concentration of 500 cells/mL in NSMB + 1 × GFs.
3. Aliquot 10 μL of cell suspension into each well of a 60-well Terasaki plate.
4. Score the number of cells present in each well at 2 h. Every well should have either 0 or 1 cell present; in rare cases, 2 cells may be visible.
5. Score the number of cells present at 24 h and 48 h; some NSPCs may have started to divide.
6. At 48 h, supplement wells that still contain at least 1 viable cell with 10 μL of NSBM + 2 × GFs.
7. Score the number of wells in which neurospheres comprising ≥4 cells have formed.

3.8 Differentiation Assay to Quantitate Neurogenic Potential

1. Prepare 8-well chamber slides by coating each well with 150 μL of Poly-D-lysine + 20 μg/mL laminin. This will ensure that NSPCs will be able to adhere.
2. Incubate at RT for 1 h.
3. Remove Poly-D-lysine + laminin and wash 2 × 5 min with PBS.
4. Dissociate 7-day culture neurospheres as per Subheading 3.2, **step 2**, and perform cell counts on dissociated cells.
5. Centrifuge cells at 700 × *g* for 7 min.
6. Resuspend cells in 500 μL NSBM + 1 × GFs supplemented with 1% fetal calf serum.
7. Plate 1 × 10^5 cells/well in 8-well chamber slides.
8. Every 2 days, remove the media and replace with fresh NSBM + 1 × GFs + 1% FCS.
9. After 7 days, fix cells with 4% paraformaldehyde in PBS for 30 min at RT.
10. Next, permeabilize the cells for 10 min with 0.1% Triton X-100 in PBS.
11. Perform blocking by incubating cells in 3% bovine serum albumin (BSA) in PBS for 30 min.
12. Incubate overnight at 4 °C with (100 μL per well) mouse anti-Tuj1 (diluted to 1:1000 in blocking solution).
13. Wash slides 2 × 2 min with PBS and incubate for 30 min at room temperature with Goat anti-Mouse 555 secondary Ab, diluted to 1:1000 in the blocking solution (100 μL per well).
14. Wash the slides 2 × 3 min in PBS and fix with 4% paraformaldehyde in PBS for 10 min at RT.

15. Wash slides for 2 × 3 min in PBS and incubate with rabbit anti-GFAP (diluted to 1:500 in blocking solution; 100 μL per well) overnight at 4 °C.
16. Wash slides 2 × 2 min in PBS and incubate at RT for 30 min with Goat anti-Rabbit 488 secondary Ab (100 μL per well), diluted to 1:1000 in the blocking solution.
17. Wash the slides 3 × 2 min in PBS before and incubate at RT for 5 min in DAPI diluted in PBS.
18. Remove the media chamber carefully with the provided chamber removal tool and scrape the glue off using a razor blade or a utility knife.
19. Slides should then be mounted in an aqueous mounting medium and left to dry overnight before imaging (Fig. 2).

3.9 Freezing Neurospheres

1. Grow neurospheres until they reach a size and density that indicates they are ready for passage.
2. Remove neurospheres from culturing vessel (e.g., flask, plate), and centrifuge at 700 × *g* for 7 min.
3. Remove supernatant and gently resuspend cells in 500 μL NSBM + 2 × GFs, containing B27 at a concentration of 1:25 rather than 1:50.
4. Transfer neurospheres to a cryovial and place on ice. Add 500 μL 20% DMSO in DMEM to cryovial and gently mix.
5. Cryovials should immediately be placed into a cryopreservation container, to ensure slow freezing. Once frozen, the samples should be taken to liquid nitrogen for storage.

3.10 Thawing Neurospheres

1. Pre-warm NSBM + 1 × GF (5 mL per sample) to 37 °C.
2. Remove cryovials from liquid nitrogen storage and place at 37 °C to thaw rapidly.
6. Immediately transfer thawed neurospheres to pre-warmed media and centrifuge at 700 × *g* for 7 min.
7. Remove supernatant and gently resuspend neurospheres in 5 mL of fresh NSBM + 1 × GF.

3.11 Retroviral Transduction

1. Plate HEK-293gp cells in 60cm^2 plates in DMEM/FCS/PS and grow until ~90% confluent.
2. Add 500 μL 2 × HEPES-buffered saline, pH 7.0 (HeBS) to a 10 mL polypropylene tube.
3. Combine 10 μg plasmid/plate with 10 μg/plate vesicular stomatitis virus glycoprotein (VSVG) envelope plasmid and 0.5 M $CaCl_2$ in a final volume of 250 μL with sterile H_2O.

4. Add plasmid/VSVG mixture dropwise to the HeBS while bubbling air through the HeBS (e.g., using a pipette boy or similar).
5. Incubate reaction mix for 20 min at RT.
6. Add dropwise to HEK-293gp cells and incubate for 24 h at 37 °C with 5%CO_2.
7. Remove media and replace with 10 mL NSBM.
8. Incubate cells for 72 h at 37 °C with 5% CO_2.
9. Remove cells from wells and centrifuge at 1400 × *g* for 5 min.
10. Filter supernatant through a 0.45 μm syringe filter to remove cellular debris.
11. Centrifuge supernatant at 25,000 × *g* for 90 min in an ultracentrifuge to concentrate the retrovirus.
12. Remove 9.5 mL supernatant and resuspend the retrovirus particles gently in 1 mL fresh NSBM. Virus can be stored at −70 °C in 500 μL aliquots.
13. Coat individual wells in a 24-well tissue culture plate with 19 μg rFN in 200 μL PBS for a minimum 16 h at 4 °C.
14. Wash each well with 400 μL PBS + 2% BSA for 30 min at RT.
15. Wash 2× with 2 mL PBS at RT.
16. Add 500 μL retroviral supernatant to each well and centrifuge the plates at 1500 × *g* for 60 min at 32 °C to facilitate binding of retroviral particles to the rFN.
17. Dissociate neurospheres as per Subheading 3.2, **step 2**, remove the viral supernatant, and plate 5 × 10^5 NSPCs in NSBM + 1 × GFs into each well (*see* **Note 4**).
18. Culture for 24 h, remove NSBM, and add 500 μL neurosphere neutralization solution.
19. Incubate for 3 min, gently triturate cells with a pipette, and transfer to a 15 mL tube.
20. Centrifuge at 1300 × *g* for 5 min.
21. Remove supernatant, resuspend cells in 2 mL NSBM + 1 × GFs, and transfer to an ultralow attachment 6-well plate (*see* **Note 5**).
22. Grow neurospheres as in Subheading 3.2.
23. If the retrovirus harbored a fluorescent marker, visualize successfully infected neurospheres using fluorescence microscopy (Fig. 3).

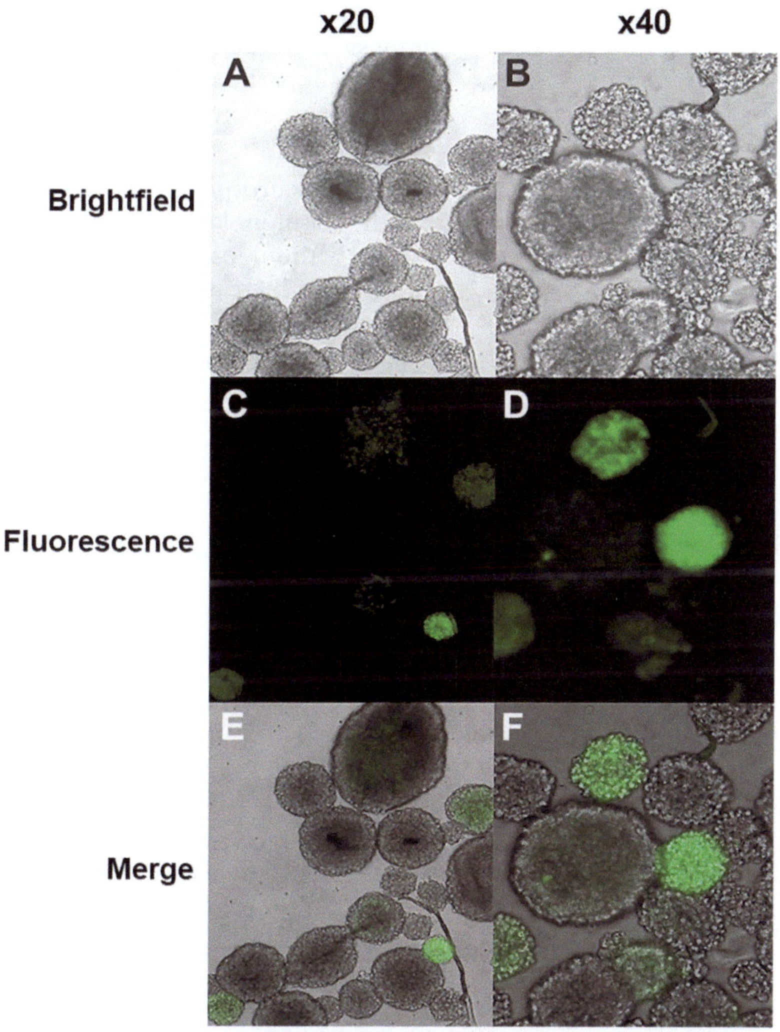

Fig. 3 Retroviral transduction of NSPCs. Neurospheres (shown here at both 20× and 40×) were grown for 7 days, disaggregated, grown as a monolayer, infected with retrovirus, and allowed to reform neurospheres. These were examined under brightfield optics (**a**, **b**) and under blue light (**c**, **d**) to identify GFP+ neurospheres. Merged photographs are also shown (**e**, **f**)

4 Notes

1. Dissect each embryo in a separate dish. If the embryos need to be genotyped, remove a small piece of limb tissue, leftover brain, or yolk sac for subsequent DNA extraction and genotyping.
2. The ganglionic eminences (GEs) can be identified because they will appear darker under brightfield microscopy than the overlying cortical layers.

3. To avoid losing neurospheres when changing half medium, gently swirl the plate to collect the neurospheres in the center of each well (particularly for neurospheres >7 days). The media can then be removed from the edges ensuring minimal neurosphere loss.
4. Plating the neurospheres on these plates allows neural stem cell growth as a monolayer culture (rather than as floating neurospheres) without inducing differentiation. A monolayer culture will greatly increase retroviral infection efficiency in this technique.
5. Using a plasmid containing a fluorescent reporter can be extremely beneficial, as positive cells and the infection efficiency can be determined using fluorescence-activated cell sorting (FACS) analysis.

References

1. Reynolds BA, Weiss S (1993) Central nervous system growth and differentiation factors: clinical horizons--truth or dare? Curr Opin Biotechnol 4(6):734–738. https://doi.org/10.1016/0958-1669(93)90058-5
2. Dworkin S, Malaterre J, Hollande F, Darcy PK, Ramsay RG, Mantamadiotis T (2009) cAMP response element binding protein is required for mouse neural progenitor cell survival and expansion. Stem Cells (Dayton, Ohio) 27(6):1347–1357. https://doi.org/10.1002/stem.56
3. Malaterre J, Mantamadiotis T, Dworkin S, Lightowler S, Yang Q, Ransome MI, Turnley AM, Nichols NR, Emambokus NR, Frampton J, Ramsay RG (2008) c-Myb is required for neural progenitor cell proliferation and maintenance of the neural stem cell niche in adult brain. Stem Cells (Dayton, Ohio) 26(1):173–181. https://doi.org/10.1634/stemcells.2007-0293
4. Dworkin S, Auden A, Partridge DD, Daglas M, Medcalf RL, Mantamadiotis T, Georgy SR, Darido C, Jane SM, Ting SB (2017) Grainyhead-like 3 (Grhl3) deficiency in brain leads to altered locomotor activity and decreased anxiety-like behaviors in aged mice. Dev Neurobiol 77(6):775–788. https://doi.org/10.1002/dneu.22469
5. Dworkin S, Mantamadiotis T (2010) Targeting CREB signalling in neurogenesis. Expert Opin Ther Targets 14(8):869–879. https://doi.org/10.1517/14728222.2010.501332
6. Luo Y, Shen H, Liu HS, Yu SJ, Reiner DJ, Harvey BK, Hoffer BJ, Yang Y, Wang Y (2013) CART peptide induces neuroregeneration in stroke rats. J Cereb Blood Flow Metab 33(2):300–310. https://doi.org/10.1038/jcbfm.2012.172
7. Natsumeda M, Maitani K, Liu Y, Miyahara H, Kaur H, Chu Q, Zhang H, Kahlert UD, Eberhart CG (2016) Targeting notch signaling and autophagy increases cytotoxicity in glioblastoma neurospheres. Brain Pathol (Zurich, Switzerland) 26(6):713–723. https://doi.org/10.1111/bpa.12343
8. D'Avanzo C, Sliwinski C, Wagner SL, Tanzi RE, Kim DY, Kovacs DM (2015) gamma-Secretase modulators reduce endogenous amyloid beta42 levels in human neural progenitor cells without altering neuronal differentiation. FASEB J 29(8):3335–3341. https://doi.org/10.1096/fj.15-271015
9. Agasse F, Roger M, Coronas V (2004) Neurogenic and intact or apoptotic non-neurogenic areas of adult brain release diffusible molecules that differentially modulate the development of subventricular zone cell cultures. Eur J Neurosci 19(6):1459–1468. https://doi.org/10.1111/j.1460-9568.2004.03259.x

Chapter 10

Testing Prepulse Inhibition of Acoustic Startle in Rodents

Maarten van den Buuse and Emily J. Jaehne

Abstract

Prepulse inhibition (PPI) is a measure of sensorimotor gating which is widely used in rodents to study information processing and attention dysfunction. PPI is commonly measured in rats and mice using automated equipment. Here, we present details of a PPI testing protocol extensively used in previous studies. The protocol includes a set pulse-alone startle level and prepulse–pulse combinations with varying interval and intensity. Variations of this protocol can be used depending on the experimental aim or equipment and software version.

Key words Behavior, Sensorimotor gating, Startle reflex, Prepulse inhibition

1 Introduction

Sensory gating and sensorimotor gating are protective mechanisms in the brain that function to gate or filter irrelevant sensory stimulation, thereby preventing sensory overstimulation and allowing for coherent thought [1–4]. Prepulse inhibition (PPI) is a measure of sensorimotor gating and is widely used in rodents to study information processing and attention dysfunction. Indeed, PPI of acoustic startle can be easily measured with relatively similar methodology in several species, including rats, mice, primates, and humans. A startle response is elicited by presentation of a short, loud acoustic stimulus (pulses or clicks) that usually induces a measurable whole-body startle response in rats and mice and a reflex eyeblink response in humans. However, when the acoustic startle stimulus is immediately preceded by a lower-intensity prestimulus (prepulse), the subsequent behavioral response is reduced (see [4–6]). This mechanism is mediated by a well-defined neural circuit, including a short brain-stem pathway mediating the startle response and several modulating brain regions, such as the hippocampus, nucleus accumbens, and frontal cortex [1, 7, 8].

Sebastian Dworkin (ed.), *Neurobiology: Methods and Protocols*, Methods in Molecular Biology, vol. 2746,
https://doi.org/10.1007/978-1-0716-3585-8_10,

PPI is deficient in a number of mental illnesses, including schizophrenia, obsessive-compulsive disorder, and attention deficit disorder, and many of these conditions are associated with deficits in brain regions that are part of the PPI regulatory circuit [3, 7, 8]. In experimental animals, deficits in PPI can be induced by acute or chronic environmental conditions or pharmacologically by treatment with dopamine receptor agonists, serotonin 1A and 2A receptor agonists, or NMDA receptor antagonists [4, 9, 10]. In addition, there is a large literature on genetically modified models with relevance to psychiatric illness that display PPI disruption [4, 9]. Several studies have shown that deficits in humans or relevant animal models can be reversed by treatment with antipsychotic drugs, such as clozapine, and PPI can be used to test novel compounds with potential antipsychotic properties [4, 11, 12].

Here, we present PPI methodology which we have used for several years in our animal models (e.g., [4, 5, 13–19]). This protocol is based upon pioneering work from Geyer, Swerdlow, and Braff (for references, see [2, 8, 9, 20]). In our PPI protocol, the startle stimuli are 115 dB in intensity and 40 ms in duration; 32 of these startle stimuli are delivered throughout the protocol, allowing calculation of the average of four blocks of eight, showing startle habituation during the PPI session [21]. Background noise is set at 65 dB for mice (70 dB for rats if used in larger startle boxes). The intensity of 20 ms prepulses is commonly expressed as dB over baseline, in this case 2, 4, 8, and 16 dB. The interval between the onset of the prepulse and the onset of the startle pulse is commonly 100 ms, but here, a 30 ms interval is also included. These different stimulus-onset asynchronies (SOAs) may produce variant results (Fig. 1), including the occurrence of negative PPI, i.e., prepulse facilitation (PPF), which may be caused by differential brain circuitry involved [1, 8, 22–24].

2 Materials

1. Experimental and control mice or rats.
2. 80% ethanol in spray bottle and paper towel.
3. Automated acoustic startle setup such as San Diego Instruments SR-LAB Startle Response System including SR-Lab software. See https://sandiegoinstruments.com/product/sr-lab-startle-response/ for images and technical details. Other commercially-available or custom-made equipment can be used to produce similar results.
4. Sink access for cleaning of animal holding cylinders.

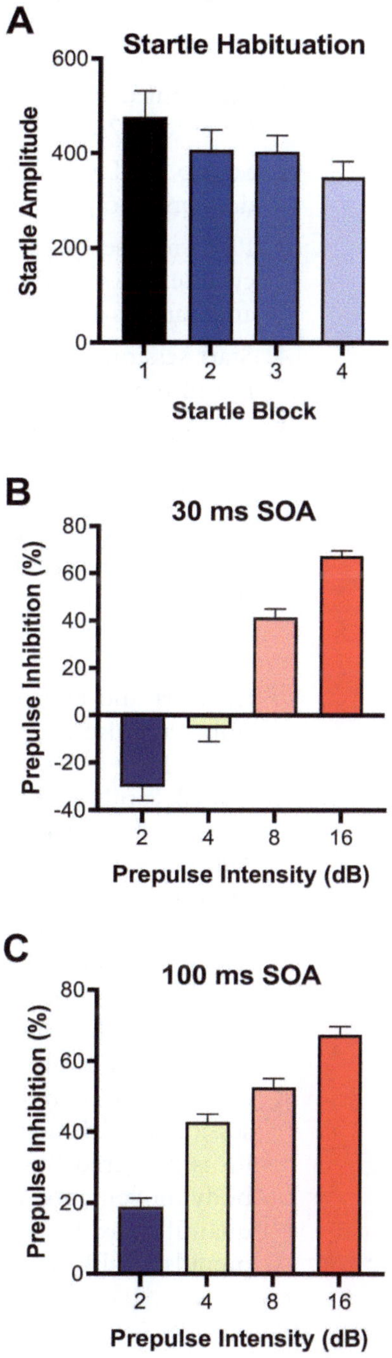

Fig. 1 Example output of data calculated from a typical PPI test session in Sprague Dawley rats (unpublished data). Panel **a** shows that startle amplitude (expressed in arbitrary units) decreases across four blocks of eight 115 dB startle-alone pulses, including one block at the start and end of the session and two throughout the session, indicating habituation to the pulse. Panels **b** and **c** show results from prepulse trials that consist of a single 115 dB pulse preceded either 30 ms (panel **b**) or 100 ms (panel **c**) by a 20 ms prepulse of

3 Methods

1. Allow mice or rats to habituate to the behavioral testing facility for at least 7 days prior to the start of any experiments.
2. Set up SR-Lab software and confirm ready to start recording data prior to placing the animal in the apparatus (*see* **Note 1**).
3. Place mouse or rat individually in appropriately sized startle cylinders in sound-attenuating chambers, and ensure all doors and barriers are closed properly (*see* **Note 2**).
4. Start session on computer immediately after placing mice and rats in the apparatus.
5. Leave for set amount of time according to previously set protocols in SR-Lab software. In our PPI protocol for mice and rats, the test session runs for approximately 45 min (Table 1), and the interval between the 104 trials is determined by a custom sequence of intertrial intervals included at the start of the programmed protocol, here varying between 12 and 24 s. Four blocks of eight 115 dB startle-alone pulses of 40 ms (here called p115) are included, including one block at the start and end of the session and two throughout the session. Prepulse trials consist of a single 115 dB pulse preceded either 100 or 30 ms SOA by a 20 ms prepulse of 2, 4, 8, or 16 dB (pp02, pp04, pp08, pp16 in Table 1) over baseline. Custom trial names reflect whether they include the 30 or 100 ms SOA and the level of prepulse intensity. For example, 30pp08p115 here means a prepulse of 8 dB over baseline, with an interval of 30 ms between its start and the onset of the 115 dB startle pulse. Each individual trial definition includes sound level settings which must be determined by prior calibration of the sound level produced by the equipment (see examples in Table 2). Throughout the session, 65 dB (mice) or 70 dB (rats) of background white noise is maintained. The session also includes eight trials during which no sound stimulus is delivered ("nostim" trials) to detect possible nonspecific body movements, for a total of 104 trials [15, 17, 18]. PPI is quantified as the difference between stimulus responses during prepulse-pulse and pulse-alone trials and expressed as a

Fig. 1 (continued) 2, 4, 8, or 16 dB over baseline background noise. PPI is quantified as the difference between stimulus responses during prepulse-pulse and pulse-alone trials and expressed as a percentage of pulse-alone responses [11–19]. PPI % is lower overall during trials with the 30 ms SOA than the 100 ms SOA and increases with increasing prepulse intensity at both intervals used. Note the occurrence of negative PPI (i.e., PPF) at the 30 ms SOA and low-intensity prepulse levels

Table 1
Example PPI protocol. The version of SR-Lab software will determine how the individual components of the protocol are programmed and entered

Session name:
Background analog level: 401 (this level is determined by prior calibration of the equipment)
No. of record samples:100
Samples per second: 1000
Intertrial interval # 1: 18
Intertrial interval # 2: 28
Intertrial interval # 3: 24
Intertrial interval # 4: 14
Intertrial interval # 5: 20
Intertrial interval # 6: 12
Intertrial interval # 7: 16
Intertrial interval # 8: 20
Trial # 1: p115-1
Trial # 2: p115-1
Trial # 3: p115-1
Trial # 4: p115-1
Trial # 5: p115-1
Trial # 6: p115-1
Trial # 7: p115-1
Trial # 8: p115-1
Trial # 9: 30pp16p115
Trial # 10: pp04p115
Trial # 11: nostim
Trial # 12: p115
Trial # 13: 30pp08p115
Trial # 14: pp02p115
Trial # 15: p115
Trial # 16: pp08p115
Trial # 17: 30pp16p115
Trial # 18: pp04p115
Trial # 19: 30pp02p115

(continued)

Table 1
(continued)

Trial # 20: 30pp08p115
Trial # 21: pp08p115
Trial # 22: p115
Trial # 23: 30pp04p115
Trial # 24: pp02p115
Trial # 25: 30pp02p115
Trial # 26: pp04p115
Trial # 27: pp16p115
Trial # 28: p115
Trial # 29: p115
Trial # 30: 30pp04p115
Trial # 31: pp16p115
Trial # 32: p115
Trial # 33: nostim
Trial # 34: 30pp16p115
Trial # 35: pp04p115
Trial # 36: p115
Trial # 37: pp16p115
Trial # 38: nostim
Trial # 39: 30pp08p115
Trial # 40: pp02p115
Trial # 41: pp08p115
Trial # 42: 30pp04p115
Trial # 43: p115
Trial # 44: pp08p115
Trial # 45: pp02p115
Trial # 46: nostim
Trial # 47: 30pp16p115
Trial # 48: pp04p115
Trial # 49: p115
Trial # 50: nostim
Trial # 51: pp08p115
Trial # 52: p115

(continued)

Table 1
(continued)

Trial # 53: 30pp02p115
Trial # 54: pp16p115
Trial # 55: p115
Trial # 56: 30pp16p115
Trial # 57: 30pp08p115
Trial # 58: p115
Trial # 59: pp16p115
Trial # 60: nostim
Trial # 61: 30pp04p115
Trial # 62: 30pp02p115
Trial # 63: pp04p115
Trial # 64: p115
Trial # 65: 30pp02p115
Trial # 66: pp02p115
Trial # 67: 30pp16p115
Trial # 68: p115
Trial # 69: pp08p115
Trial # 70: 30pp04p115
Trial # 71: 30pp16p115
Trial # 72: p115
Trial # 73: nostim
Trial # 74: pp16p115
Trial # 75: p115
Trial # 76: 30pp08p115
Trial # 77: 30pp04p115
Trial # 78: pp02p115
Trial # 79: pp08p115
Trial # 80: nostim
Trial # 81: 30pp08p115
Trial # 82: 30pp02p115
Trial # 83: pp04p115
Trial # 84: 30pp16p115
Trial # 85: pp02p115

(continued)

Table 1
(continued)

Trial # 86: pp16p115
Trial # 87: 30pp08p115
Trial # 88: pp02p115
Trial # 89: 30pp04p115
Trial # 90: 30pp08p115
Trial # 91: pp04p115
Trial # 92: 30pp02p115
Trial # 93: pp16p115
Trial # 94: 30pp04p115
Trial # 95: 30pp02p115
Trial # 96: pp08p115
Trial # 97: p115-4
Trial # 98: p115-4
Trial # 99: p115-4
Trial # 100: p115-4
Trial # 101: p115-4
Trial # 102: p115-4
Trial # 103: p115-4
Trial # 104: p115-4
Sequence repetitions:1
Acclimation period (mins):3

percentage of pulse-alone responses (see Fig. 1, e.g., of typical results). We have developed an Excel spreadsheet which sorts the raw data from the PPI equipment and calculates relevant averages and percentages. Some of the trial naming in the protocol in Table 1 is to facilitate sorting of the data from each trial type.

6. At end of session, inspect the mouse or rat for any health concerns (*see* **Note 3**) before returning to their home cage. Return all animals to their housing room at the end of the daily testing period.
7. Ensure all equipment is cleaned with 80% ethanol solution or other specified solutions, prior to testing of the first animal and in between subsequent animals.

Table 2
Trial definitions used in the PPI protocol in Table 1. Note the analog levels mentioned in these trial definitions must be determined by calibration of the SR-Lab equipment and software as per procedures provided by the supplier. The version of SR-Lab software will determine how the individual components of the trial definitions are programmed and entered

Trial name: 30pp02p115
At 0 milliseconds: analog level, 415
At 0 milliseconds: wait length (ms), 20
At 20 milliseconds: background
At 20 milliseconds: wait length (ms), 10
At 30 milliseconds: analog level, 704
At 30 milliseconds: record data
At 30 milliseconds: wait length (ms), 40
At 70 milliseconds: background
At 70 milliseconds: end of trial
Trial name: 30pp04p115
At 0 milliseconds: analog level, 428
At 0 milliseconds: wait length (ms), 20
At 20 milliseconds: background
At 20 milliseconds: wait length (ms), 10
At 30 milliseconds: analog level, 704
At 30 milliseconds: record data
At 30 milliseconds: wait length (ms), 40
At 70 milliseconds: background
At 70 milliseconds: end of trial
Trial name: 30pp08p115
At 0 milliseconds: analog level, 455
At 0 milliseconds: wait length (ms), 20
At 20 milliseconds: background
At 20 milliseconds: wait length (ms), 10
At 30 milliseconds: analog level, 704
At 30 milliseconds: record data
At 30 milliseconds: wait length (ms), 40
At 70 milliseconds: background
At 70 milliseconds: end of trial

(continued)

Table 2
(continued)

Trial name: 30pp16p115
At 0 milliseconds: analog level, 509
At 0 milliseconds: wait length (ms), 20
At 20 milliseconds: background
At 20 milliseconds: wait length (ms), 10
At 30 milliseconds: analog level, 704
At 30 milliseconds: record data
At 30 milliseconds: wait length (ms), 40
At 70 milliseconds: background
At 70 milliseconds: end of trial
Trial name: pp02p115
At 0 milliseconds: analog level, 415
At 0 milliseconds: wait length (ms), 20
At 20 milliseconds: background
At 20 milliseconds: wait length (ms), 80
At 100 milliseconds: analog level, 704
At 100 milliseconds: record data
At 100 milliseconds: wait length (ms), 40
At 140 milliseconds: background
At 140 milliseconds: end of trial
Trial name: pp04p115
At 0 milliseconds: analog level, 428
At 0 milliseconds: wait length (ms), 20
At 20 milliseconds: background
At 20 milliseconds: wait length (ms), 80
At 100 milliseconds: analog level, 704
At 100 milliseconds: record data
At 100 milliseconds: wait length (ms), 40
At 140 milliseconds: background
At 140 milliseconds: end of trial
Trial name: pp08p115
At 0 milliseconds: analog level, 455
At 0 milliseconds: wait length (ms), 20

(continued)

Table 2
(continued)

At 20 milliseconds: background
At 20 milliseconds: wait length (ms), 80
At 100 milliseconds: analog level, 704
At 100 milliseconds: record data
At 100 milliseconds: wait length (ms), 40
At 140 milliseconds: background
At 140 milliseconds: end of trial
Trial name: pp12p115
At 0 milliseconds: analog level, 482
At 0 milliseconds: wait length (ms), 20
At 20 milliseconds: background
At 20 milliseconds: wait length (ms), 80
At 100 milliseconds: analog level, 704
At 100 milliseconds: record data
At 100 milliseconds: wait length (ms), 40
At 140 milliseconds: background
At 140 milliseconds: end of trial
Trial name: pp16p115
At 0 milliseconds: analog level, 509
At 0 milliseconds: wait length (ms), 20
At 20 milliseconds: background
At 20 milliseconds: wait length (ms), 80
At 100 milliseconds: analog level, 704
At 100 milliseconds: record data
At 100 milliseconds: wait length (ms), 40
At 140 milliseconds: background
At 140 milliseconds: end of trial
Trial name: p115
At 0 milliseconds: analog level, 704
At 0 milliseconds: record data
At 0 milliseconds: wait length (ms), 40
At 40 milliseconds: background
At 40 milliseconds: end of trial

(continued)

Table 2 (continued)

Trial name: nostim
At 0 milliseconds: record data
At 0 milliseconds: wait length (ms), 10
At 10 milliseconds: end of trial

8. Following each session, cylinders can be removed from the chambers and rinsed with water to clean. Ensure not to get any electrical components wet.
9. Analyze data generated by SR-Lab software. Startle response should be recorded for each stimulus presented and %PPI calculated (Fig. 1).

4 Notes

1. When using any behavioral monitoring software, always keep any eye on the data being collected during behavioral testing, and make a note if it looks like there were any problems, for example, startle responses not changing with different prepulse intensities. Checking connections and rearranging cables can help fix problems with incorrect data collection. Equipment with tone and movement detection settings such as used for PPI should be calibrated regularly to ensure correct parameters are used.
2. Mice and rats to be injected with compounds before PPI should receive injection immediately prior to placing in chamber or a set amount of time depending on the pharmacokinetics of the drug and time of action.
3. Provided all steps are followed correctly, you should not expect any negative health outcomes following behavioral testing. However, excess porphyrin around the eyes and nose may appear due to stress.

References

1. Gómez-Nieto R, Hormigo S, López DE (2020) Prepulse inhibition of the auditory startle reflex assessment as a hallmark of brainstem sensorimotor gating mechanisms. Brain Sci 10:639
2. Light GA, Braff DL (1999) Human and animal studies of schizophrenia-related gating deficits. Curr Psychiatry Rep 1:31–40
3. van den Buuse M (2007) Pre-pulse inhibition. In: Fink G (ed) Encyclopedia of stress, 2nd edn. Academic Press, Oxford, pp 180–183
4. van den Buuse M (2010) Modeling the positive symptoms of schizophrenia in genetically modified mice: pharmacology and methodology aspects. Schizophr Bull 36:246–270

5. Gogos A, van den Buuse M (2022) Sex differences in psychosis: focus on animal models. Curr Top Behav Neurosci (in press)
6. van den Buuse M, Garner B, Gogos A, Kusljic S (2005) Importance of animal models in schizophrenia research. Aust N Z J Psychiatry 39: 550–557
7. Koch M (1999) The neurobiology of startle. Prog Neurobiol 59:107–128
8. Swerdlow NR, Geyer MA, Braff DL (2001) Neural circuit regulation of prepulse inhibition of startle in the rat: current knowledge and future challenges. Psychopharmacology (Berl) 156:194–215
9. Geyer MA, McIlwain KL, Paylor R (2002) Mouse genetic models for prepulse inhibition: an early review. Mol Psychiatry 7:1039–1053
10. Van den Buuse M, Garner B, Koch M (2003) Neurodevelopmental animal models of schizophrenia: effects on prepulse inhibition. Curr Mol Med 3:459–471
11. Gray L, van den Buuse M, Scarr E, Dean B, Hannan AJ (2009) Clozapine reverses schizophrenia-related behaviours in the metabotropic glutamate receptor 5 knockout mouse: association with N-methyl-D-aspartic acid receptor up-regulation. Int J Neuropsychopharmacol 12:45–60
12. McOmish CE, Burrows E, Howard M, Scarr E, Kim D, Shin HS, Dean B, van den Buuse M, Hannan AJ (2008) Phospholipase C-beta1 knockout mice exhibit endophenotypes modeling schizophrenia which are rescued by environmental enrichment and clozapine administration. Mol Psychiatry 13:661–672
13. Gururajan A, Hill RA, van den Buuse M (2015) Brain-derived neurotrophic factor heterozygous mutant rats show selective cognitive changes and vulnerability to chronic corticosterone treatment. Neuroscience 284:297–310
14. Jaehne EJ, Ameti D, Paiva T, van den Buuse M (2017) Investigating the role of serotonin in methamphetamine psychosis: unaltered behavioral effects of chronic methamphetamine in 5-HT_{1A} knockout mice. Front Psych 8:61
15. Jaehne EJ, Chong EMS, Sbisa A, Gillespie B, Hill R, Gogos A, van den Buuse M (2021) TrkB agonist 7,8-dihydroxyflavone reverses an induced prepulse inhibition deficit selectively in maternal immune activation offspring: implications for schizophrenia. Behav Pharmacol 32: 404–412
16. Jaehne EJ, Kent JN, Antolasic EJ, Wright BJ, Spiers JG, Creutzberg KC, De Rosa F, Riva MA, Sortwell CE, Collier TJ, van den Buuse M (2022) Behavioral phenotyping of a rat model of the BDNF Val66Met polymorphism reveals selective impairment of fear memory. Transl Psychiatry 12:93
17. Manning EE, van den Buuse M (2013) BDNF deficiency and young-adult methamphetamine induce sex-specific effects on prepulse inhibition regulation. Front Cell Neurosci 7:92
18. Notaras MJ, Hill RA, Gogos JA, van den Buuse M (2017) BDNF Val66Met genotype interacts with a history of simulated stress exposure to regulate sensorimotor gating and startle reactivity. Schizophr Bull 43:665–672
19. van den Buuse M, Ruimschotel E, Martin S, Risbrough VB, Halberstadt AL (2011) Enhanced effects of amphetamine but reduced effects of the hallucinogen, 5-MeO-DMT, on locomotor activity in 5-HT_{1A} receptor knockout mice: implications for schizophrenia. Neuropharmacology 61:209–216
20. Swerdlow NR, Weber M, Qu Y, Light GA, Braff DL (2008) Realistic expectations of prepulse inhibition in translational models for schizophrenia research. Psychopharmacology (Berl) 199:331–388
21. Abel K, Waikar M, Pedro B, Hemsley D, Geyer M (1998) Repeated testing of prepulse inhibition and habituation of the startle reflex: a study in healthy human controls. J Psychopharmacol 12:330–337
22. Aasen I, Kolli L, Kumari V (2005) Sex effects in prepulse inhibition and facilitation of the acoustic startle response: implications for pharmacological and treatment studies. J Psychopharmacol 19:39–45
23. Naysmith LF, Williams SCR, Kumari V (2022) The influence of stimulus onset asynchrony, task order, sex and hormonal contraception on prepulse inhibition and prepulse facilitation: methodological considerations for drug and imaging research. J Psychopharmacol 36: 1234–1242
24. Wynn JK, Dawson ME, Schell AM, McGee M, Salveson D, Green MF (2004) Prepulse facilitation and prepulse inhibition in schizophrenia patients and their unaffected siblings. Biol Psychiatry 55:518–523

Chapter 11

Syngeneic Mouse Model of Glioblastoma: Intracranial Implantation of GL261 Cells

Paulina Kaminska, Salwador Cyranowski, Paulina Pilanc, and Anna R. Malik

Abstract

Glioblastoma (GBM) is the most aggressive and prevalent primary brain malignancy in adults. Current treatments provide limited benefit, and thus, the median overall survival of GBM patients is only 15 months. GBM progression is highly dependent on its ability to evade immune response, so understanding the mechanisms behind GBM-driven immunosuppression seems crucial for designing more efficient therapies. Animal models of GBM constitute a convenient tool in glioma research, and several different approaches have been already developed to model this disease in vivo, including genetic and xenograft models. Here, we describe a murine syngeneic model of glioma which recapitulates many of the key features of human disease, including complex tumor microenvironment. We present an optimized protocol for stereotactic intracranial implantation of GL261 cells into C57BL/6 mice which results in tumor growth in the striatum. This model has been widely used to get insight into glioma biology, as well as in the studies aiming at the development and validation of new therapeutic approaches.

Key words Glioblastoma, Intracranial implantation, Stereotactic surgery, GL261, Mouse model

1 Introduction

Gliomas are tumors of the central nervous system (CNS) classified by the World Health Organization (WHO) into several types, based on their morphological and molecular features [1]. Glioblastoma (GBM) is the most common and aggressive primary brain malignancy, for which the standard of care, consisting of surgical resection combined with radio- and chemotherapy, is notoriously failing. Due to highly diffusive growth and strong immunosuppressive microenvironment of the tumor, treatment of GBM patients remains palliative with median overall survival of 15 months [2, 3]. Experimental murine models of GBM that recapitulate human disease are commonly used in the development of new

Sebastian Dworkin (ed.), *Neurobiology: Methods and Protocols*, Methods in Molecular Biology, vol. 2746,
https://doi.org/10.1007/978-1-0716-3585-8_11,

therapeutic strategies which are urgently needed to improve patient survival and quality of their life. Murine models are also widely used in basic science research studying molecular mechanisms which drive tumor progression [4].

In vivo murine models of GBM in general can be divided into three categories [5, 6]:

1. Genetically engineered mouse models. Genome manipulations on the level of somatic or germline cells allow investigating the impact of genetic alterations on tumor progression and treatment efficiency.
2. Xenografts of human glioma cultured cells or patient-derived tumor cells. Immunodeficient mice can undergo intracranial implantations with glioma cell lines of human origin or patient's tissue. These models are widely used for studying the microenvironment of human GBM and mechanisms of treatment resistance.
3. Syngeneic models of GBM. Intracranial implantation of mouse glioma cells into immunocompetent mice with intact immune system and similar genetic background reduces risk of transplantation rejection. Such models are commonly used in preclinical studies and investigation of therapies involving the response of the immune system.

Implantations of glioma cells described in points 2 and 3 can be also performed subcutaneously; however, this method lacks the CNS microenvironment, which is essential in glioma pathology [7]. Due to this fact, we are focusing on the description of intracranial implantations.

None of the above-mentioned models fully recapitulate the features of human GBM. These models vary in invasiveness, immunogenicity, and rate of tumor progression. It is crucial to choose an in vivo model of GBM which is most suitable for the goals of the study [8]. Here, we demonstrate the protocol for precise stereotactic implantation of glioma cells (GL261) in the syngeneic murine model of glioma, which results in growth of the tumor recapitulating certain clinical features of GBM. Although this protocol is written for GL261 cells, it can be used for the intracranial implantation of various glioma cells; however, the number of cells implanted depends significantly on the cell line of choice.

The GL261 cell line was established over 20 years ago and was generated by inducing brain tumor in a C57BL/6 mouse after treatment with chemical carcinogen, methylcholanthrene, and maintained by a series of syngeneic transplantations [9]. Tumors which develop after intracranial implantation of GL261 cells share many histological features with human GBM including tumor necrosis, pseudopalisades, neovascularization, invasion, hypercellularity, and inflammation [10]. They also recapitulate some features of the immunosuppressive microenvironment [11]. What is more, genetically modified GL261 cells which express the luciferase gene are commonly used for intracranial implantations due to the possibility of in vivo bioluminescent imaging of the tumor which allows monitoring its growth and response to treatments [12]. Taking into account its easy and stable culturing in vitro, GL261 cell line constitutes a convenient model to study molecular mechanisms driving GBM progression.

Please note that animal research is a subject to regulations in many countries. It is thus crucial to follow the local law in this respect. As intracranial implantations are harmful for animals, the procedure should be considered in the context of the 3Rs principle—reduction, replacement and refinement, which is essential in securing laboratory animals' well-being [13].

2 Materials

2.1 Cell Culture

1. GL261 cells (*see* **Note 1**).
2. T75cm^2 cell culture flasks (*see* **Note 2**).
3. Dulbecco's Modified Eagle's Medium (DMEM) (1 g/L glucose).
4. Complete cell culture medium: DMEM supplemented with 10% fetal bovine serum (FBS) and 1% penicillin–streptomycin.
5. Trypsin–EDTA (0.25%).
6. Cell counting equipment: hemocytometer or automatic cell counter.
7. Cell culture incubator maintaining constant temperature, humidity, and carbon dioxide (CO_2) concentration.
8. Centrifuge with temperature regulation.
9. 15 mL conical tubes.
10. 0.5 mL (or smaller) microcentrifuge tubes (*see* **Note 3**).
11. Serological pipettes.
12. Automatic pipettes.

2.2 Surgery Room Equipment

Surgery room should be easy to keep clean and to sanitize. It should not be used for purposes other than animal surgeries.

1. UV lamp.
2. Halogenated anesthetic supply system.
3. Halogenated anesthetic charcoal scavenger (e.g., Harvard Apparatus Fluosorber®).

4. Digital mouse stereotaxic instrument with warming base or heating pad (*see* **Note 4**).
5. Microinjector syringe pump.
6. Surgical lamp.

2.3 Consumables and Surgical Tools

All surgical tools, gauze, and cotton swabs should be sterilized prior to the procedure. Analgesics, liquids, surgical drapes, and syringes should be kept in the surgery room and opened on-site only after its sterilization with UV light and disinfection with appropriate disinfectant, e.g., Velox®.

1. Medical-grade ethanol for skin and tool disinfection.
2. Hydrogen peroxide—H_2O_2 (3%).
3. Ø 60 mm plastic Petri dishes (*see* **Note 5**).
4. Disposable surgical drapes.
5. Scissors, trimmer, or commercially available depilatory cream—depending on the operator's preference.
6. Lacrimal fluid substitute (e.g., Vidisic® gel). Lacrimal fluid substitute prevents intraoperative drying of the cornea.
7. Small pieces of gauze. Cut it into small squares prior to the procedure.
8. Hemostatic sponge.
9. Plastic, disposable 1 mL syringes with 22G needles.
10. Systemic analgesics:
    - Opioid analgesic: butorphanol (e.g., Butomidor®);
    - Nonsteroid anti-inflammatory drug: meloxicam (e.g., Metacam®).
11. Local analgesics:
    - Bupivacaine (e.g., Marcaine®);
    - 4% lidocaine cream.
12. Sterile 0.9% NaCl for injections.
13. Sterile water for injections.
14. Cotton swabs (preferably with long handles).
15. Permanent marker used only for skull marking.
16. Electric micromotor handheld drill.
17. Metal Ø 1–1.2 mm drill bit.
18. Glass, precision 1 μL syringe (e.g., Hamilton® syringe).
19. Additional cage with fresh bedding and water-dampened "chow" for animals recovering after surgery.
20. Timer.

21. USP 5–0 nylon nonabsorbable sutures (e.g., Atramat®) with 3/8 circle 16 mm needle.
22. Surgical wax (bone wax) for local bone hemostasis.
23. Surgical tools: forceps, curved serrated iris forceps, mosquito forceps, bent tip tweezers, penfield dissector, serrated tweezers (*see* **Note 6**).
24. Disposable curved scalpel blades (no. 10, 11).
25. Scale.

3 Methods

3.1 Preparation of GL261 Cells

Complete cell culture medium, PBS, and trypsin–EDTA (0.25%) should be pre-warmed to 37 °C.

DMEM used for cell preparation on a day of intracranial implantations should be prechilled to 4 °C.

1. Thaw out GL261 cells in advance so they are subcultured at least two times prior to the intracranial implantation (10–12 days before). Culture cells at 5% CO_2 and 37 °C on T75cm^2 cell culture flask in complete cell culture medium.
2. On the day of the procedure, cells should be sub-confluent (*see* **Note 7**).
3. Remove cell medium and gently rinse cells with 10 mL of PBS.
4. Add 1 ml of trypsin–EDTA (0.25%) and incubate at 37 °C until cells are fully detached from the flask (3–4 min).
5. Add 6 mL of fresh cell medium to the flask. Pipette thoroughly to obtain single cell suspension. Do not foam the cell suspension.
6. Transfer cells to a fresh 15 mL conical tube.
7. Determine the amount of the viable cells.
8. Transfer 1×10^6 of cells to the fresh 15 mL conical tube and centrifuge (5 min, 150×*g*, 4 °C).
9. After centrifugation, while performing the following steps, keep cells on ice.
10. Remove the supernatant carefully and discard.
11. Resuspend the pellet in 5 mL of fresh, cold DMEM.
12. Repeat centrifugation as in **step 8**.
13. Remove the supernatant carefully so there is no DMEM left above the pellet. Discard the supernatant. If there are any left-overs, use an automatic pipette to remove as much of the medium as possible without aspiring the cells.
14. Resuspend the cell pellet in 12.5 μL of fresh, cold DMEM to achieve the desired concentration of 8×10^4 cells/μL.

15. Transfer cell suspension to the microcentrifuge tube. Keep the cells on ice. Cells can be safely used for intracranial implantations up to 4 h (*see* **Note 8**).

3.2 Intracranial Implantation of GL261 Cells

The whole procedure should be performed in aseptic conditions. Make sure that the operation room was sterilized with UV light at least 1 h prior to the procedure.

1. Weigh a mouse to establish the preimplantation body mass (*see* **Note 9**).
2. Prepare drug working solutions directly in 1 mL syringes. First, aspirate the drug stock solution and next 0.9% NaCl for injections:
   - 0.5 mg/mL of Butomidor
   - 0.5 mg/mL of Metacam
   - 0.25% of Marcaine.
3. Calculate the amount of each drug which will be administered to the animal according to its body mass:
   - Butomidor: 2 mg/kg.
   - Metacam: 2 mg/kg.
   - Marcaine: 8 mg/kg.
4. Anesthetize the animal in an isoflurane chamber using 4% isoflurane delivered in oxygen-enriched air (95% oxygen; flow rate 1 L/min).
5. Attentively observe the mouse's behavior and allow it to aspirate isoflurane for 1 min from the moment it appears to be sedated, (i.e., does not move around the cage and its respiratory rate is lowered).
6. Put the mouse aside and pinch its toe using serrated tweezers to make sure it's not responding to external stimuli. If so, place the mouse again in the chamber until the efficient anesthesia is achieved.
7. Shave the top of the animal's head using scissors, trimmer, or depilatory cream.
8. Place the animal in a stereotactic instrument. Immobilize its head by hooking the mouse's upper incisors in the hole of the tooth bar and placing ear bars at the entry of the ear canal on each side (Fig. 1).
9. Tight the nose clamp over the snout and ensure the head stability by gently applying pressure with a cotton swab or a gloved finger.
10. Put on the snout mask with isoflurane supply. At the beginning, keep 2% of isoflurane concentration. Oxygen flow rate should stand at 1 L/min.

Fig. 1 A mouse placed in a stereotactic instrument

11. Confirm that the mouse is sufficiently anesthetized by assessing the animal's spontaneous response to a toe pinch. If the animal responds to the toe pinch, delay the procedure until it completely does not react to the stimuli.
12. Lower the isoflurane concentration to the level of 1.5%.
13. Apply lacrimal fluid substitute on the animal's eyes.
14. Put small patches of sterile gauze on the top of the gel droplets to protect the eyes from excess light.
15. Administer analgesics subcutaneously:
    - Butomidor and Metacam: two opposite flanks of the animal's body.
    - Marcaine: head at the site of surgery.
16. Apply ethanol solution to the top of the mouse's head with cotton swabs to disinfect the incision site.
17. One more time assesses the animal's spontaneous response to a toe pinch to make sure it does not react to external stimuli.
18. Make a longitudinal 1 cm incision from the front toward the back of the head with a scalpel. Use cotton swabs to move the folds of the skin and expose the skull.
19. Rub the skull with cotton swabs soaked in H_2O_2 until the cranial sutures are visible.

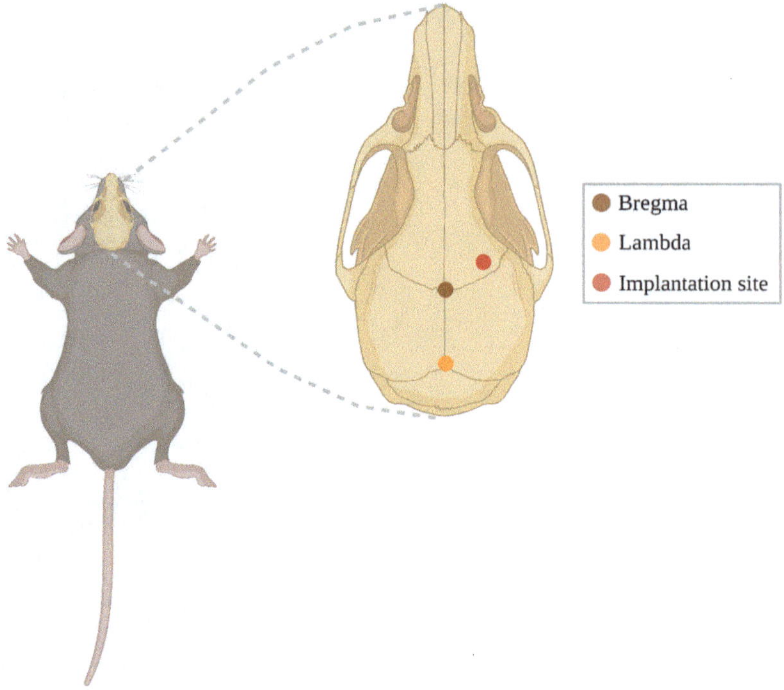

Fig. 2 Scheme of a murine skull indicating locations of bregma, lambda, and the implantation site

20. Mount the 1 μL glass syringe onto the arm of the stereotactic instrument connected to the microinjector syringe pump.
21. Locate the bregma (Fig. 2) and set the needle so it points it. Make sure you did not confuse bregma with lambda. Zero out the X, Y, and Z coordinates on the digital stereotactic display.
22. According to bregma as a zero point, adjust the stereotaxic coordinates to the following values:

    Z: −1.0 mm (anterior–posterior).

    X: +2.0 mm (medial–lateral).

    Such coordinates direct the syringe to the right hemisphere.
23. Lower the syringe so that the needle is hovering above the skull. Mark the place pointed by the needle on the skull with a permanent marker. Lift the syringe.
24. Use a drill to make a Ø1–1.2 mm hole in the skull where the point was marked. If bleeding occurs, use a hemostatic sponge to stop it.
25. Dry the hole using a cotton swab.
26. Check if the drilling was performed in the desired site by lowering the syringe over the hole and controlling values of stereotaxic coordinates.

27. To resuspend the cells, gently vortex or flick the microcentrifuge tube.
28. Take up 1 μL of cell suspension to the 1 μL glass syringe using the microinjector syringe pump. Control the level of suspension in the syringe so it doesn't exceed 1 μL. Avoid introducing air bubbles.
29. Using dry cotton swabs, remove any of the cell suspension fluid that appears at the tip of the needle to prevent contamination of the incision site with glioma cells which may result in tumor growing in the extracranial space.
30. Lower the syringe so that the needle is aligned with the brain surface. Zero out Y (dorsal–ventral) coordinate.
31. Slowly lower the tip of the needle until it reaches a depth of −3.5 mm. This coordinate allows implantation of cells into the striatum.
32. Wait for 1 min.
33. Lift the syringe so Y coordinate is −3.0 mm (*see* **Note 10**).
34. Allow the needle to stay positioned at Y for 1 min. Assess the site of injection for the presence of blood or translucent fluid (cerebrospinal fluid). Dry the area if needed.
35. Set the microinjector syringe pump to deliver 0.25 μL of the cell suspension per 1 min.
36. Start injecting the cell suspension. It should last 4 min.
37. After administration, wait for 2 min (*see* **Note 11**).
38. Start lifting the syringe at the rate of 1 mm per 1 min. In total, it should last 3 min.
39. Dry the hole using a cotton swab.
40. Take the 1 μL glass syringe out from the stereotactic device, and wash it in water (aspirate sterile injection-grade water and empty the syringe several times). Make sure that any blood or tissue that sticks to the needle is removed by gently rubbing the wet needle against a paper towel.
41. Take a piece of surgical wax using a penfield dissector and apply it onto the opening in the skull.
42. Using curved tweezers, approximate the wound edges so that the wound is closed. Suture the incision with 3–4 over and over interrupted stitches.
43. Mark the animal so it's distinguishable from other animals in the cage.
44. Put lidocaine cream on the stitches using a cotton swab.
45. Remove the animal from the stereotactic instrument and place it in a box on a heating blanket set to 37 °C until it recovers. Monitor the animal's response after anesthesia.

46. Transfer mouse to a new cage and ensure easy access to food and water (*see* **Note 12**).
47. When all animals from one cage are already implanted with GL261 cells, transfer them to their home cage.
48. Monitor the welfare of mice regularly until the end point of your experiment (*see* **Note 13**).
49. Sacrifice the mice at the time point of your choice (*see* **Notes 14** and **15**).

4 Notes

1. Depending on the kind of experiment you want to perform, you can use different kinds of GL261 cells, e.g., with luciferase or GFP/tdTomato genes. Prior to culturing the cells, check if they require specific antibiotic selection.
2. In our experiments, we use T75cm^2 cell culture flasks which allow to get 3–4 × 10^6 of GL261 cells from one flask. However, it is also possible to culture cells using, e.g., Ø 100 mm plastic Petri dishes, suitable for cells.
3. 0.2–0.5 mL tubes are small enough to be handled between the needle's tip and the animal's scalp.
4. If you do not have access to a digital mouse stereotaxic instrument, you can operate it manually. The warming base of the instrument is important for keeping the animal's body temperature at physiological level during surgery. Inhalatory anesthesia can cause hypothermia.
5. Petri dishes are necessary to ensure easy access to ethanol and H_2O_2 during surgery. Pour liquids into two separate Petri dishes, cover, and label correctly.
6. In between individual surgeries, surgical tools should be disinfected using medical-grade ethanol or sterilized in hot bead sterilizer (250 °C for 15 s). The latter have to be followed by cooling down the tools prior to contacting tissues.
7. GL261 cells start to detach from the cell culture flask when they cover ~75% of its surface. Do not use cells which reached a higher percentage of confluency. Cells which will be implanted into the murine brain have to be in the growth phase. Overgrowing can change the properties of cells, which can affect tumor growth.
8. Time needed for the implantation of one mouse is 45–55 min, and it is not advised to use GL261 cells 4 h after the moment of preparation. Take into account how many animals you are going to implant and plan the number of necessary flasks/Petri dishes with the cells.

9. For intracranial implantations of GL261 cells, we usually use 8–12-week-old C57BL/6 mice with body mass between 20 and 30 g.
10. The initial level of the needle's tip (Y, −3.5 mm) is necessary to create a 0.5 mm space for the cells which are then implanted from the starting point of Y, −3.0 mm.
11. It is important to lift the syringe at the indicated rate as it prevents backward outflow of the cell suspension.
12. Do not transfer mice after implantation immediately to a home cage, as other animals which have not been operated yet might become aggressive and hurt the operated mouse.
13. Animals' welfare has to be carefully monitored after the procedure. Any signs of distress, pain, or substantial weight loss should be considered as humane end point criteria.
14. The presented protocol allows to maintain tumor growth in animals up to approx. 25 days. However, keep in mind that the time point of animal sacrifice depends on individual experimental aims and the topic of a study. You may need to adjust the time point of sacrifice to your experimental needs. Please note that over the course of tumor growth animals begin to develop neurological symptoms and at critical time point (around day 20 postsurgery) begin to abruptly lose body weight (Fig. 3).
15. Implantation of different number of GL261 cells using described protocol will affect overall survival of animals.

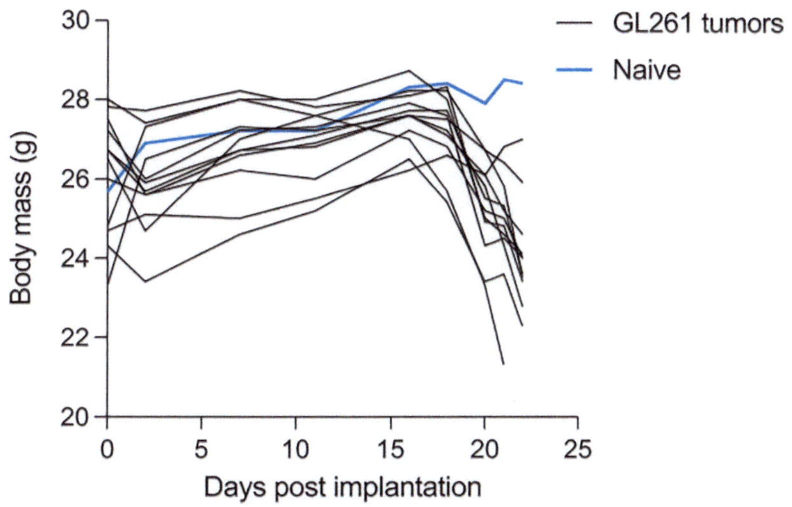

Fig. 3 Body mass of mice implanted with GL261 cells. Mice were implanted with GL261-GFP cells according to the protocol described here. Each line represents body mass changes of individual mouse after the surgery. Naive—mouse that did not undergo implantation

References

1. Louis DN, Perry A, Wesseling P et al (2021) The 2021 WHO classification of tumors of the central nervous system: a summary. Neuro-Oncology 23:1231–1251
2. Pilanc P, Wojnicki K, Roura AJ et al (2021) A novel oral arginase 1/2 inhibitor enhances the antitumor effect of PD-1 inhibition in murine experimental gliomas by altering the immunosuppressive environment. Front Oncol 11: 703465
3. Gieryng A, Pszczolkowska D, Walentynowicz KA et al (2017) Immune microenvironment of gliomas. Lab Investig 97:498–518
4. Jacobs VL, Valdes PA, Hickey WF, de Leo JA (2011) Current review of in vivo GBM rodent models: emphasis on the CNS-1 tumor model. ASN Neuro 3:171–181
5. Haddad AF, Young JS, Amara D et al (2021) Mouse models of glioblastoma for the evaluation of novel therapeutic strategies. Neurooncol Adv 3(1):vdab100
6. Jin F, Jin-Lee HJ, Johnson AJ (2021) Mouse models of experimental glioblastoma. In: Gliomas. Exon Publications, Brisbane, pp 15–46
7. Lenting K, Verhaak R, ter Laan M et al (2017) Glioma: experimental models and reality. Acta Neuropathol 133:263–282
8. Szatmári T, Lumniczky K, Désaknai S et al (2006) Detailed characterization of the mouse glioma 261 tumor model for experimental glioblastoma therapy. Cancer Sci 97:546–553
9. Maes W, van Gool SW (2011) Experimental immunotherapy for malignant glioma: lessons from two decades of research in the GL261 model. Cancer Immunol Immunother 60: 153–160
10. Abdelwahab MG, Sankar T, Preul MC, Scheck AC (2011) Intracranial implantation with subsequent 3D in vivo bioluminescent imaging of murine gliomas. J Vis Exp 57:e3403
11. Oh T, Fakurnejad S, Sayegh ET et al (2014) Immunocompetent murine models for the study of glioblastoma immunotherapy. J Transl Med 12:107
12. Shelton LM, Mukherjee P, Huysentruyt LC et al (2010) A novel pre-clinical in vivo mouse model for malignant brain tumor growth and invasion. J Neuro-Oncol 99:165–176
13. Wange RL, Brown PC, Davis-Bruno KL (2021) Implementation of the principles of the 3Rs of animal testing at CDER: past, present and future. Regul Toxicol Pharmacol 123: 104953

Chapter 12

Measurement of Murine Neuromuscular Function Using the In Situ Preparation

S. T. C. Kucewicz, M. Borkowski, J. E. Church, and C. van der Poel

Abstract

The presence and progression of a neuromuscular pathology can impact on the contractile force production of a muscle. Hence, measurements of force production can be an important tool for the evaluation of disease progression. In this chapter, we describe how to perform in situ function testing on the *tibialis anterior* muscle using a murine model. Performing neuromuscular in situ function testing allows force measurements to be recorded in a physiologically relevant environment.

Key words Neuromuscular performance, Force measurement, Twitch force, Tetanic force, Hindlimb force, Tibialis anterior muscle, In situ function

1 Introduction

The skeletal muscle is a highly metabolically active tissue which is consistently adapting its structure and metabolism due to functional demands. Several neuromuscular pathologies such as muscular dystrophies and amyotrophic lateral sclerosis (ALS) can affect skeletal muscle homeostasis, resulting in dysfunction and atrophy [1–3]. In situ neuromuscular function testing is a technique that can be used to quantify the functional performance of skeletal muscle by determining the force that is directly produced by a specific skeletal muscle. Unlike ex vivo function measurements, in situ contractile force measurements are performed on skeletal muscle that remains intact in the body, with complete nerve and blood supply, and therefore allow measurements to occur in a more physiologically relevant environment [4]. For in situ function of the mouse hindlimb, the proximal end of the *tibialis anterior* (TA) muscle remains connected to the body, while the distal end is ligated to the lever arm of a dual-mode lever or force transducer. TA muscles can be stimulated to produce contractile force by

Sebastian Dworkin (ed.), *Neurobiology: Methods and Protocols*, Methods in Molecular Biology, vol. 2746, https://doi.org/10.1007/978-1-0716-3585-8_12,

electrodes that are either touching the innervating common peroneal nerve or via surface electrodes placed directly onto the surface of the TA muscle. In normal and healthy TA muscles, both direct stimulation and nerve stimulation result in the same profile of force production. In the case of neuromuscular or neurodegenerative pathologies, a combination of direct stimulation and nerve stimulation protocols could help identify whether function deficits are due to nerve or neuromuscular junction (NMJ) issues or to changes in the muscle itself [5].

In this chapter, we describe how to assess murine neuromuscular in situ function of the TA skeletal muscle using both direct and nerve stimulation techniques. The protocol begins with a single isolated twitch protocol, followed by a force–frequency protocol.

2 Materials

1. Krebs-Ringer solution: 137 mM NaCl, 24 mM $NaHCO_3$, 11 mM D-glucose, 5 mM KCl, 1 mM $NaH_2PO_4H_2O$, and 1 mM $MgSO_4$ in ultrapure. Bubble solution with carbogen (5% CO_2 in O_2) for 10 min. Check pH: The pH of the solution should be 7.4; adjust if necessary. Add 2 mM CaCl2 to the solution. Check and readjust pH if necessary. Add 0.025 mM d-tubocurarine chloride into the solution. Check and readjust pH if necessary. Heat the solution to 37 °C.
2. 5.0 silk braided suture.
3. Electric hair clippers.
4. Sodium pentobarbitone (6 mg/mL).
5. Dissecting plate.
6. Heat mat or heating lamp.
7. Infrared skin thermometer.
8. Pasteur pipette.
9. Vernier calipers.
10. 16-gauge needle.
11. Scalpel.
12. Aurora Scientific 1300A 3-in-1 whole animal system.

3 Methods

The following methods are to be performed under standard laboratory conditions unless stated otherwise.

3.1 Preparation of Mouse Hindlimb

1. Anesthetize the mouse with an intraperitoneal injection of sodium pentobarbitone (6 mg/mL). Determine the depth of sedation by performing a toe pinch (pedal reflex) on the mouse. If adequate sedation has occurred, then no reaction should be observed to the toe pinch (*see* **Note 1**).
2. Place mouse onto a 37 °C heat mat or place a 37 °C heat lamp above the mouse to help maintain its body temperature. Throughout the procedure, body temperature should be monitored with an infrared thermometer to ensure hypothermia does not occur.
3. Prepare the skin by shaving the desired hindlimb of the mouse with an electric shaver (*see* **Note 2**).
4. Place the mouse in a supine position, fixing the leg to a secure block. This can be achieved using either an elastic band around a metal block or a dissecting pin through the foot into a cork mat. Make an incision into the skin of the hindlimb ensuring the underlying muscle tissue remains untouched. Using blunt dissection techniques, peel back the skin to expose skeletal muscles of the anterior portion of the lower hindlimb, exposing the TA as well as the proximal anterolateral upper hindlimb to expose the tensor fasciae latae (TFL) and biceps femoris (BF) muscles (Fig. 1).

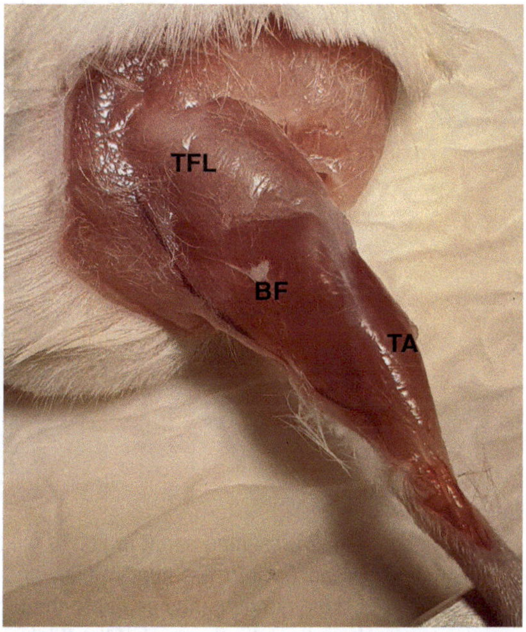

Fig. 1 Exposed skeletal muscles of the mouse hindlimb. In preparation for in situ testing, the skin of the hindlimb should be removed to expose the skeletal muscles of the anterior portion of the lower hindlimb. The TA should be visible, along with the proximal anterolateral upper hindlimb comprising of the tensor fasciae latae (TFL) and biceps femoris (BF) muscles. These landmarks are important for identifying where the sciatic nerve is located

5. Using a stereomicroscope, locate the distal tendon of the muscle (*see* **Note 3**). Using 5.0 silk braided suture, tie a square (reef) knot, followed by a loop knot around the distal tendon. This suture will be used to attach the muscle to the lever arm of the Aurora Scientific 1300A 3-in-1 whole animal system.
6. Sever the distal tendon of the muscle inferiorly to the site of the knot, leaving as much of the tendon as possible. Gently dissect the TA muscle from surrounding tissue by detaching any fascia covering the TA muscle. Continue to dissect 1/3 to ½ of the length of the TA from surrounding musculature and tibial bone. Using the distal tendon, the TA should be able to be carefully lifted from the body.
7. The biceps femoris anterior and posterior muscles are gently separated from the lateral aspect of the thigh to get access to the sciatic nerve (Fig. 2). Gently clear the surrounding connective tissue around the sciatic nerve, and follow the nerve

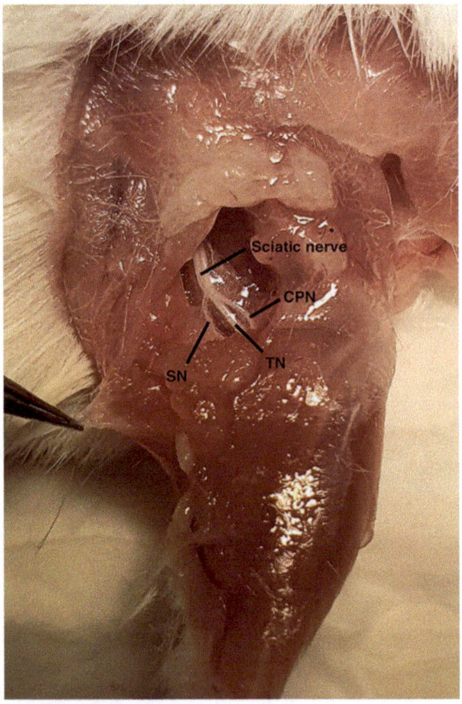

Fig. 2 Dissected upper region of the hindlimb exposing the sciatic nerve where it branches into the sural, tibial, and common peroneal nerves. To expose the sciatic nerve, gently separate the biceps femoris anterior and posterior muscles from the lateral aspect of the thigh. The sciatic nerve branches into three nerves: sural nerve (SN), tibial nerve (TN), and common peroneal nerve (CPN). The CPN innervates the anterior lower limb; therefore, it will be the stimulating nerve being used for the in situ function testing of the TA muscle

proximal to the hip, to where the sciatic nerve branches into the sural, tibial, and common peroneal nerves. The tibial and sural nerves innervate the posterior region of the lower limb, whereas the common peroneal nerve (CPN) innervates muscles of the anterior lower limb. While performing muscle and nerve isolation, superfuse the exposed muscles and nerve with Krebs-Ringer solution (*see* **Note 4**).

8. Place the animal in a prone position on the animal platform heated to 37 °C, or place a heat lamp over the animal to help ensure the maintenance of body temperature.
9. Secure the animal to the setup using a pin (*see* **Note 5**) and foot clamp. If required, fix the position of the mouse using additional pins. Using the suture tied to the distal tendon, attach the TA to the lever arm (Fig. 3a).
10. For stimulation via the nerve, position the stimulating electrodes under the CPN preventing other muscles being activated during the contraction process (Fig. 3b). For direct muscle stimulation, insert the fine needle electrodes into the mid-belly of the TA muscle (Fig. 3c; *see* **Note 6**). If performing both nerve stimulation and direct muscle stimulation, perform the nerve stimulation first followed by the direct muscle stimulation protocol.
11. Once the mouse, muscle, and electrodes are in position, ensure the knot is secured by performing a 100 Hz tetanic contraction.

3.2 Determination of Optimal Length (L_o) of the Muscle

1. Prior to performing any neuromuscular function measurements, optimal muscle length (L_o) must be determined. Optimal muscle length is defined as the muscle length that produces the maximum relative twitch tension (Pt@Lo). To achieve this, adjust the muscle length using the micropositioner [6]. After each adjustment, stimulate the TA muscle with a single electrical pulse (1 Hz) to elicit a twitch response. When performing the individual twitch, use the following parameters: initial delay 0.2 s, pulse width 0.2 ms. Note the total force produced (peak force minus resting, passive baseline), and allow at least 30 sec rest between each stimulation. As the muscle length is increased, peak twitch force will increase, along with the passive baseline force. When the difference of the peak and baseline twitch force no longer increases, peak twitch at optimal muscle length (L_o) has been reached.
2. Using vernier calipers, record L_o, measuring tendon to tendon, and routinely check throughout the procedure to ensure L_o is maintained.

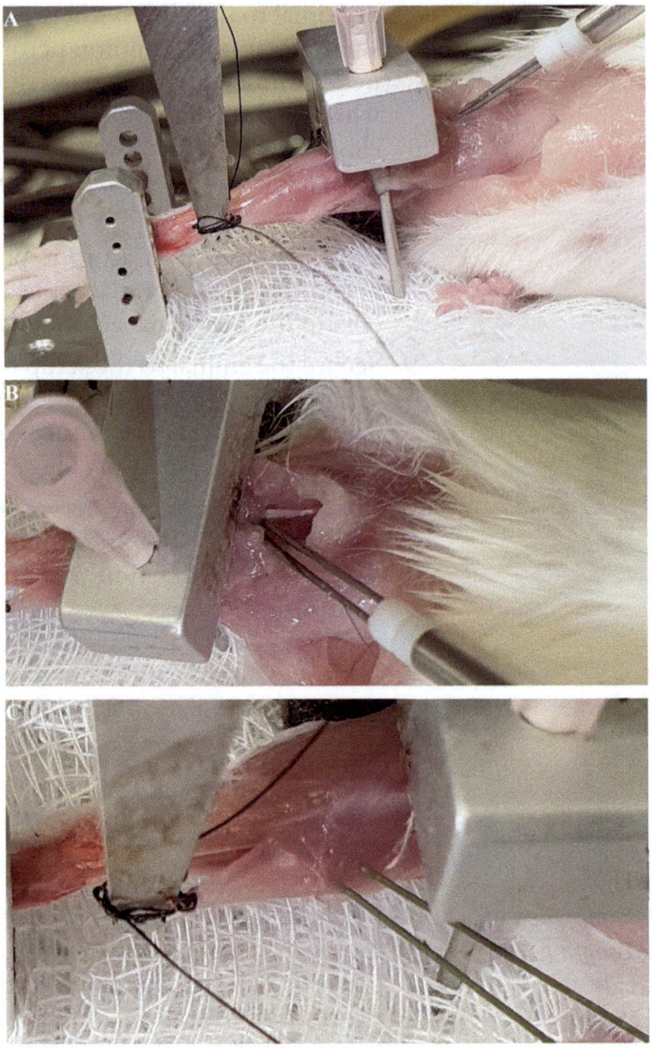

Fig. 3 Setup of the mouse hindlimb on the Aurora Scientific 1300A 3-in-1 whole animal system for both direct and nerve stimulation. (**a**) The foot should be in a fixed position, secured using a foot clamp. The distal tendon of the TA muscle is attached to the lever arm using the 5.0 silk braided suture. A pin should be placed through the patella tendon to ensure there is limited movement at the knee joint. (**b**) For nerve stimulation, the stimulating electrode is be positioned under the CPN to prevent other muscles from being activated during the contraction process. (**c**) For direct muscle stimulation, the fine needle electrodes are placed into the mid-belly of the TA muscle

3.3 Determination of Maximal Isometric Force (P_o) Using a Force–Frequency Relationship

1. Allow a 2 min rest between the determination of optimal muscle length and the beginning of the force–frequency protocol.
2. Ensure muscle length is still set to the previously recorded L_o and that exposed muscle and nerve are superfused with the Krebs-Ringer solution.

3. The muscle preparation should be stimulated every 2 min at the following increasing pulse frequencies, 10, 20, 30, 40, 50, 80, 100, 150, 200, and 250 Hz with a 0.2 s initial delay, 0.2 ms pulse width, and 0.2–0.35 s stimulus duration.
4. The maximum absolute isometric tetanic force (P_o) is determined from the maximum force produced at the plateau of the frequency–force relationship minus the resting passive baseline.
5. If performing both nerve stimulation and direct muscle stimulation, repeat this protocol using direct muscle stimulation.

3.4 Dissection of TA Muscle

1. Carefully dissect the TA muscle from the body, detaching the muscle from the proximal tendon.
2. Trim any remaining tendon and blot the muscle dry with filter paper. Record the weight of the muscle (*see* **Note** 7).
3. If you would like to perform further histomorphological analysis on the muscle, freeze the muscle using thawing isopentane and store the muscle at a temperature of −80 °C [7].

3.5 Post-function Calculations

1. Optimal muscle fiber length (L_f) is determined by multiplying L_o by the TA muscle length to fiber length ratio of 0.6 [8].
2. The total cross-sectional area (CSA) of the muscle is approximated using measurements collected through the procedure. To approximate CSA, use the following equation [9]:

$$\mathrm{CSA} = \frac{\text{Muscle weight (g)}}{[(L_o(\mathrm{mm}) * 0.6) * 1.06(\mathrm{mg\,mm^{-3}})]}$$

   Where 1.06 mg mm^{-3} is the density of mammalian skeletal muscle and optimal muscle length is multiplied by 0.6 which is the known TA optimal fiber length L_f/L_o ratio [8].
3. Specific isometric tetanic force (sP_o; kN/m^2) can be calculated by dividing absolute isometric tetanic force (P_o) by the calculated total muscle CSA (*see* **Note 8**).

4 Notes

1. If at any point the mouse appears to show signs of regaining consciousness, a supplemental intraperitoneal injection of 10% of the initial administered dose should be given to keep the animal at the appropriate depth of anesthesia.
2. Hair removal cream can also be applied at this stage to remove all hair from the hindlimb. Removal of hair makes preparation and post-function tissue harvest easier.
3. If you cannot locate the distal tendon, you may be required to dissect a greater area of skin toward the region where the

tendon is located. Once located, gently separate the tendon from any surrounding fascia; this will make it easier to tie a knot around the tendon.

4. During the preparation and throughout the experiment, it is important to monitor the exposed tissues and prevent these tissues from becoming too dry. To prevent this from occurring, every 4 min, superfuse the exposed muscle tissue with Krebs-Ringer solution using a Pasteur pipette or equivalent.
5. Secure and fix the hindlimb by placing a pin through the patella tendon.
6. For both procedures, it is important to stabilize the electrodes in a fixed position to prevent the electrodes changing position and influencing muscle contraction. Taping the leads of the electrodes to the Aurora Scientific 1300A 3-in-1 platform is possible, as is using an adjustable soldering stand with alligator clips to hold the electrodes.
7. It is important to record muscle weight as the size of the muscle can affect its function. The muscle weight is also important for obtaining other post-protocol measurements.
8. Since absolute isometric tetanic force (P_o) is affected by total muscle CSA, it is best practice to normalize P_o to total muscle CSA.

References

1. Loeffler JP, Picchiarelli G, Dupuis L, Gonzalez De Aguilar JL (2016) The role of skeletal muscle in amyotrophic lateral sclerosis. Brain Pathol 26(2):227–236
2. Gerlinger-Romero F, Addinsall AB, Lovering RM, Foletta VC, van der Poel C, Della-Gatta PA, Russell AP (2019) Non-invasive assessment of dorsiflexor muscle function in mice. J Vis Exp 143. https://doi.org/10.3791/58696
3. Rahimov F, Kunkel LM (2013) Cellular and molecular mechanisms underlying muscular dystrophy. J Cell Biol 201(4):499–510
4. Croes SA, von Bartheld CS (2007) Measurement of contractile force of skeletal and extraocular muscles: effects of blood supply, muscle size and in situ or in vitro preparation. J Neurosci Methods 166(1):53–65
5. Pratt SJP, Shah SB, Ward CW, Inacio MP, Stains JP, Lovering RM (2012) Effects of in vivo injury on the neuromuscular junction in healthy and dystrophic muscles. Physiol J 591(2):559–570
6. Barker R, Horvath D, Van der Poel C, Murphy R (2017) Benefits of prenatal taurine supplementation in preventing the onset of acute damage in the mdx mouse. PLoS Curr. https://doi.org/10.1371/currents.md.9a3e357a0154d01050b591601cbd4fdb
7. Lovering RM, Roche JA, Goodall MH, Clark BB, McMillan A (2011) An in vivo rodent model of contraction-induced injury and non-invasive monitoring of recovery. J Vis Exp 51. https://doi.org/10.3791/2782
8. Burkholder TJ, Fingado B, Baron S, Lieber RL (1994) Relationship between muscle fiber types and sizes and muscle architectural properties in the mouse hindlimb. J Morphol 221(2):177–190
9. Lynch GS, Hinkle RT, Chamberlain JS, Brooks SV, Faulkner JA (2001) Force and power output of fast and slow skeletal muscles from mdx mice 6-28 months old. J Physiol 535(Pt 2):591–600

Chapter 13

Cryosectioning and Immunohistochemistry Using Frozen Adult Murine Brain Neural Tissue

Jemma Gasperoni and Sebastian Dworkin

Abstract

Cryopreservation and immunohistochemistry offer a comprehensive, robust, and simple methodology to investigate neural patterning and cellular function. Rapid freezing of the whole brain allows excellent preservation of neural ultrastructure and tissue architecture without destroying sensitive protein epitopes that are often compromised following standard paraffin embedding histological techniques. Here, we present a rapid and simple protocol for employing cryosectioning and subsequent immunohistochemistry in the study of adult murine brain neural tissue.

Key words Cryopreservation, Cryosectioning, Immunohistochemistry, Mouse, Brain

1 Introduction

The study of the structure and function of individual nerve cells within the brain has been a longstanding pursuit in the field of neuroscience, with the ultimate goal of understanding the complex processes that govern brain function, development, and disease [1]. Cryopreservation and sectioning to examine neural structure, followed by subsequent immunohistochemistry to detect protein expression (neural function), are two powerful techniques that have significantly advanced our ability to investigate the cellular and molecular mechanisms underlying fundamental neurobiology [2, 3].

Immunohistochemistry (IHC) combines the principles of histology and immunology to visualize the distribution and localization of specific antigens within a tissue [4]. This is achieved by using antibodies that specifically bind to the target antigen, followed by a detection system that produces a visible signal, such as a colored precipitate or fluorescence [5]. IHC has become an indispensable tool in neuroscience research, as it allows for the identification and

Sebastian Dworkin (ed.), *Neurobiology: Methods and Protocols*, Methods in Molecular Biology, vol. 2746, https://doi.org/10.1007/978-1-0716-3585-8_13,

characterization of various cell types, as well as the investigation of cellular processes and signalling pathways in the context of the intact tissue [2].

Cryopreservation and sectioning offer several advantages over other techniques for studying neural structure and function. First, the preservation of antigenicity in frozen tissue allows for the detection of a wide range of antigens, including those that may be sensitive to the chemical fixation and processing steps required for paraffin-embedded tissue sections or other histological methods [6]. Secondly, the rapid freezing of the tissue helps to minimize the effects of autolysis and other postmortem changes that can compromise the quality of the tissue and can lead to the loss or degradation of key antigens of interest. In contrast, techniques such as electron microscopy [7] or Golgi-Cox staining [8], while providing high-resolution images of neural cell structures such as soma, axons, and dendrites, do not preserve the molecular composition of the tissue as effectively as cryosectioning.

Third, the use of cryosectioning and immunohistochemistry allows for the investigation of dynamic cellular processes and signalling pathways in their native context. This is particularly valuable in the study of neurodevelopment, neurodegeneration, and other processes that involve complex interactions between different cell types and molecules [9]. Alternative techniques, such as in vitro cell culture [10, 11] or dissociated neuronal preparations, may not fully recapitulate the intricate cellular environment and interactions present in the intact brain tissue.

In conclusion, cryosectioning and IHC using adult mouse brain provide a powerful and versatile approach for investigating the cellular and molecular processes that govern brain function. As our understanding of the brain continues to grow, particularly in understanding neural connectivity and signalling between disparate brain regions, cryosectioning and IHC will undoubtedly continue to play a central role in advancing our knowledge of novel neuronal signalling pathways in the etiology of behavior, mental health, disease, aging, and degeneration.

2 Materials

1. Phosphate-buffered saline (PBS): For a 10× solution, add 80 g NaCl, 2 g KCl, 14.4 g Na_2HPO_4 to 800 mL of distilled water. Adjust to pH 7.4 and fill to 1 L. For 1× PBS, add 100 mL of 10× PBS to 900 mL of distilled water.
2. 10% (w/v) sucrose in PBS.
3. 30% (w/v) sucrose in PBS.
4. Bovine serum albumin (BSA).

5. Triton X-100 detergent.
6. 4% (w/v) paraformaldehyde (PFA) in PBS: Combine 4 g PFA with 70 mL PBS, place onto an orbital shaker (*see* **Note 1**), and mix until the PFA dissolves and the liquid becomes clear. Filter PFA solution through filter paper and add deionized water to a final volume of 100 mL.
7. PBS + 4% bovine serum albumin (BSA) + 0.1% Triton X-100 (PBS/BSA/Tx): Combine 10 g BSA, 250 μL Triton-X 100, and 25 mL 10× PBS and add deionized water to a final volume of 250 mL (*see* **Note 2**).
8. Normal goat serum blocking solution: Combine 500 μL 1 M lysine, 250 μL NGS, and 4.25 mL PBS/BSA/Tx.
9. PBS/0.1% Tx: Combine 100 mL 10 × PBS and 1 mL Triton-X, stir gently, and slowly add deionized water to a final volume of 1 L (*see* **Note 3**).
10. Sodium citrate: Dissolve 2.941 g tri sodium citrate ($Na_3C_6H_5O_7$) in 500 mL deionized water. Adjust pH to 6.0 add deionized water to a final volume of 1 L.
11. Tris–EDTA: Dissolve 1.21 g Tris(hydroxymethyl)-aminomethane ($(HOCH_2)_3CNH_2$; Tris) and 0.37 g ethylene-diaminetetraacetic acid ($[CH_2N(CH_2CO_2H)_2]_2$; EDTA) in 500 mL deionized water. Adjust pH to 9.1 and add deionized water to a final volume of 1L.
12. Isopentane (2-methylbutane).
13. Dry ice.
14. 1M lysine.
15. Normal goat serum.
16. Primary antibody: For example, to identify intermediate progenitor cells (rat anti-EOMES).
17. Secondary antibody: For example, goat anti-rat secondary with conjugated fluorophore.
18. Optimal cutting temperature compound (10.24% polyvinyl alcohol, 4.26% polyethylene glycol, 85.5% nonreactive ingredients; OCT).
19. Antigen retrieval buffer 1: Sodium citrate (pH 6): 0.1 M sodium citrate dihydrate.
20. Antigen retrieval buffer 2: Tris–EDTA (pH 9.1): 10 mM Tris, 1 mM EDTA.
21. DAPI (4′,6-diamidino-2-phenylindole).
22. Glass microscope slides and coverslips.
23. Mountant for coverslipping.

3 Methods

3.1 Cardiac Perfusion

1. Euthanize adult mice according to approved institutional animal ethics protocol. (*see* **Note 4**).
2. Transcardially perfuse using 4% (w/v) paraformaldehyde (PFA) in PBS until all blood is flushed out.
3. Set up the perfusion machine with fresh PBS and ensure the tubing is clean, free of air bubbles and primed with PBS.
4. Make an incision through the abdomen, the length of the diaphragm. With sharp scissors, cut through the connective tissue at the bottom of diaphragm to allow access to the rib cage. Be careful to avoid cutting the liver.
5. Grab the skin at the sternum and pull upward, then pierce the diaphragm with sharp scissors (pulling the skin upward before cutting ensures that the heart or major vessels will not be cut), and cut away the diaphragm laterally to the left and right.
6. Cut through each side of the rib cage being careful not to puncture the lung or heart (*see* **Note 5**).
7. Place the animal on an elevated rack with a container underneath to collect the blood and paraformaldehyde waste.
8. Insert a needle and syringe into the protrusion of the left ventricle to extend straight up about 2–3 mm. Be careful not to extend the needle too far in, as it can pierce an interior wall and compromise circulation of solutions. Do not insert the needle into the right ventricle as this will lead to the respiratory tract.
9. Make a cut in the right atrium with sharp scissors, and switch on the pump. Make sure the solution is flowing freely. If fluid is not flowing freely or is coming from animal's nostrils or mouth, reposition the needle.
10. Once cleared (*see* **Note 6**), switch the solution to 4% (w/v) paraformaldehyde (PFA) in PBS.
11. Perfuse the animal for 4–5 min or until the animal has stiffened.

3.2 Brain Extraction

1. Dissect out brain and rinse in PBS.
2. Immediately place the entire brain in 4% PFA/PBS and fix for 24 h at 4 °C.
3. Remove the brain from PBS/PFA and place in a solution of 10% sucrose in PBS at 4 °C in a 15 mL tube, until the tissue has sunk to the bottom of the tube (*see* **Note 7**).

4. Replace the 10% (w/v) sucrose solution with 30% (w/v) sucrose in PBS and incubate the whole brain at 4 °C until the tissue has once again sunk to the bottom of the tube.
5. In order to freeze the brain for sectioning, first chill the isopentane on dry ice for 30 min.
6. For whole brains or large sections, directly lower the entire tissue into isopentane. For smaller regions of the brain, such as olfactory bulbs or cerebellum, place tissue into a cryomold surrounded by OCT, before lowering into the isopentane (*see* **Note 8**).
7. Leave tissues in isopentane until frozen—tissues will need a minimum of 20 s in isopentane.
8. Once frozen, store tissues at −80 °C.

3.3 Cryosectioning

1. Place the desired section of the brain to be sectioned in OCT in a cryomold.
2. Set the cryostat to −22 (*see* **Note 9**).
3. Before the OCT is taken out of the cryomold, place a mark on the OCT to ensure that when it is removed from the mold, the tissue orientation can be determined.
4. Once the sample has been removed from the mold, place in cryostat for at least 30 min to adjust to the temperature.
5. Meanwhile, to prepare the chuck (specimen plate) with a layer of OCT (forming a platform for the sample to sit on), put layer of OCT on the chuck, and use the weighted heat extractor to flatten the OCT.
6. Cut the OCT until there is an evenly flat surface for the sample to be attached (Fig. 1).
7. Place a small amount of OCT onto the chuck and push the sample into the OCT until it is flushed with the flat surface underneath. Use a small amount of OCT to paint up the side of the sample to ensure it is secured to the chuck.
8. Leave the sample another 10 min for the OCT to dry before beginning sectioning.

3.4 Immunohistochemistry

DAY 1

1. Bring the tissue sections to room temperature (~30 min).
3. For antigen retrieval (*see* **Note 10**), boil 100 mL of antigen retrieval buffer 1 and immerse the slides immediately in the solution.
4. Incubate slides for 10 min with the lid on.
5. Boil another 100 mL of antigen retrieval buffer 1; once the first incubation period is complete, immerse slides into freshly boiled solution for a further 10 min with the lid on.

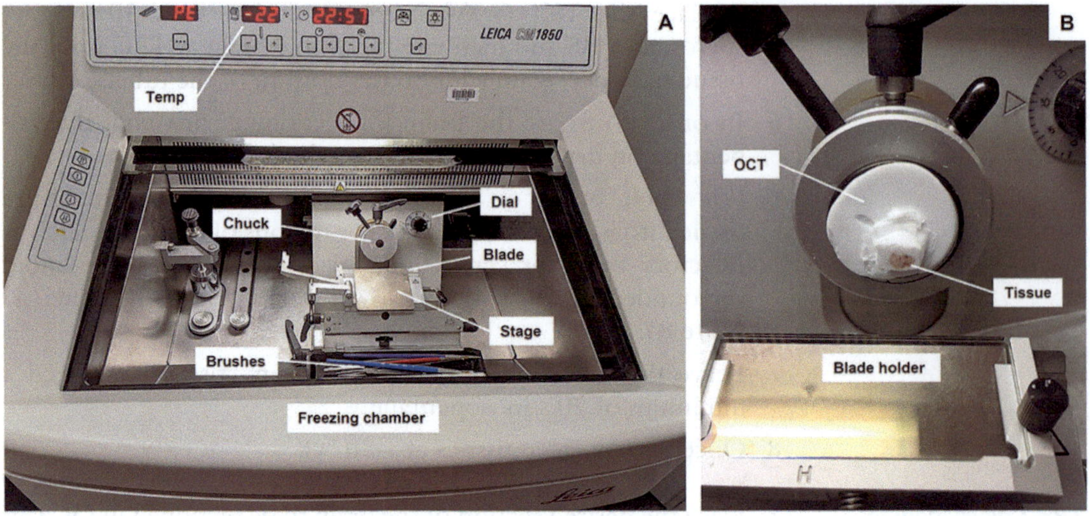

Fig. 1 The freezing chamber of a cryostat. (**a**) A typical temperature for cutting cryopreserved sections is −22 °C. The sample is held in place on the chuck, where it can be moved toward the blade using the positioning dial, to allow the user to cut sections of desired thickness. Once cut, the thin sections can be removed from the stage and onto a microscope slide using fine-haired brushes. (**b**) The tissue is frozen within the cryomold and, once extracted, is adhered to OCT on the chuck. This then allows the tissue block to be moved downward to the blade (placed in the blade holder) to cut the cryosections

6. Remove the lid and allow solution to cool for 20 min at room temperature.
7. Wash slides in PBS for 3 × 5 min.
8. Incubate the sections for 1 h in blocking solution at 4 °C.
9. Incubate the sections overnight at 4 °C with the primary antibody in PBS/BSA/TX (*see* **Note 11**).

DAY 2

10. Wash 3 × 5 min in PBS/0.1% Tx (*see* **Note 12**).
11. Incubate the slides for 1 h at 4 °C with the secondary antibody in PBS/BSA/Tx (Fig. 2).
12. Wash 3 × 5 min in PBS/0.1% Tx (*see* **Note 13**).
13. Apply DAPI diluted in PBS (*see* **Note 14**).
14. Wash the sections 3 × 5 min in PBS.
15. Gently tap the slide on the bench to get rid of excess PBS.
16. Mount with fluorescence suitable mountant (*see* **Note 15**).
17. Apply 2–3 drops and coverslip.
18. Allow to cure overnight (in the dark if using immunofluorescence) before imaging.

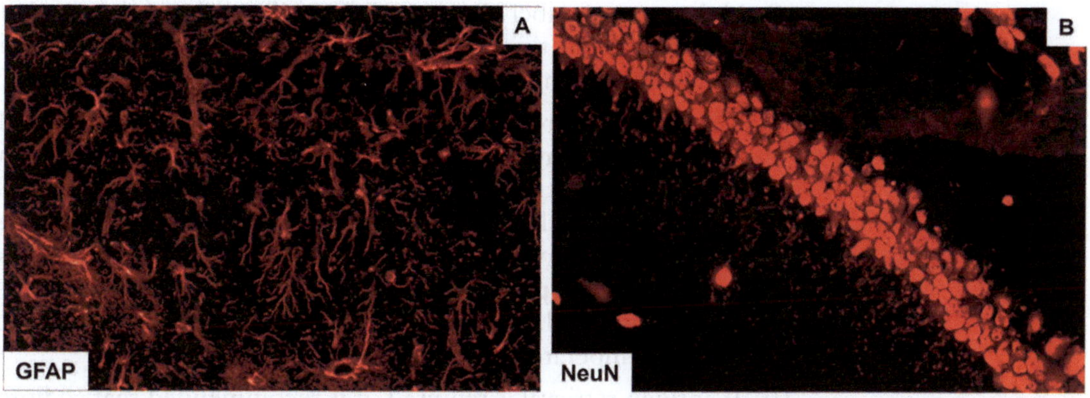

Fig. 2 Example of neuronal immunofluorescence staining of cryopreserved tissue. Using rabbit anti-mouse primary antibodies to detect (**a**) glial fibrillary acidic protein (GFAP) in astrocytes of the hippocampus and (**b**) NeuN+ neurons within the CA1 region of the hippocampus. In both cases, the secondary antibody used was donkey anti-rabbit Alexa Fluor 594

4 Notes

1. Using a heated orbital shaker will speed up this process.
2. This can be aliquoted into 10–15 mL and stored at −20 °C. Make sure that it is fully thawed before use.
3. If deionized water is added too quickly, it will cause bubbles to form, making it difficult to accurately measure.
4. For brain IHC, we would recommend cervical dislocation or CO_2 asphyxiation to reduce risks of causing any damage to the brain tissue.
5. Heart damage should not happen as long as the scissors are pressed against the rib cage on the side of the animal. Incision should be from the bottom to the top of the thoracic cavity. Also, be careful not to cut the blood vessels immediately lateral to the sternum. The rib cage should be able to be lifted such that a working environment is established with the maximal field of view. For mouse dissections, the rib cage can be cut at the level of the manubrium of the sternum and removed. The heart and lung will be in full view.
6. A good indication that the animal has been perfused adequately is if the liver is cleared. A color change from a dark red/brown color to a lighter brown color will become apparent once the blood has been cleared from the liver.
7. The time it takes for the tissue to sink will depend on both the tissue type and size. These steps will protect the tissue from damage in subsequent cryopreservation steps. This is important as during the freezing step, if not properly cryoprotected, ice crystals may form in the tissue and damage important histoarchitecture.

8. Make sure to label the cryomold well and note down what orientation the tissue is in. Using cryomolds and OCT will also work for embryonic brain tissue.
9. This temperature may vary depending on the cryostat and tissue on the day. Start at −22 °C, and then, depending on the behavior of the sample, troubleshooting may be necessary. For example, if the sample is sticking too much, then lowering the temperature would be advised. If the sample is brittle, then the cryostat temperature should be raised. Each time the temperature is raised, allow 20–30 min for the cryostat to reach the desired temperature.
10. If antigen retrieval is required, it is recommended that sodium citrate is trialled first. Tris–EDTA buffer (antigen retrieval buffer 2) is harsh and should only be used if other methods have not resulted in optimal staining.
11. The antibody dilution will vary based on the recommended/tested dilution for the specific antibody. Make sure to investigate this well in advance from either the suppliers' catalogue or published literature. It is best to use an antibody that has been referenced previously using the same tissue processing method and tissue type.
12. The rest of the protocol should be performed in a dark room. If this is not possible, then covering the slides in foil and limiting the amount of light will suffice.
13. Increase to 6 × 5 min washes if there is too much background.
14. If DAPI is already included in the mountant, then continue to **step 12**.
15. The type of mountant to use will depend on the type of staining being performed. For immunofluorescence, make sure to use an aqueous anti-fade mountant that will help prolong the fluorescent signal. For DAB staining, a mountant mixture of distyrene, a plasticizer, and xylene (DPX) should be used.

References

1. Delcambre GH, Liu JJ, Herrington JM, Vallario K, Long MT (2016) Immunohistochemistry for the detection of neural and inflammatory cells in equine brain tissue. Peerj 4:ARTN e1601. https://doi.org/10.7717/peerj.1601
2. Dworkin S, Auden A, Partridge DD, Daglas M, Medcalf RL, Mantamadiotis T, Georgy SR, Darido C, Jane SM, Ting SB (2017) Grainyhead-like 3 (Grhl3) deficiency in brain leads to altered locomotor activity and decreased anxiety-like behaviors in aged mice. Dev Neurobiol 77(6):775–788. https://doi.org/10.1002/dneu.22469
3. Juma AR, Hall NE, Wong J, Gasperoni JG, Watanabe Y, Sahota A, Damdimopoulou PE, Grommen SVH, De Groef B (2018) PLAG1 expression and target genes in the hypothalamo-pituitary system in male mice. Mol Cell Endocrinol 478:77–83. https://doi.org/10.1016/j.mce.2018.07.009
4. Vazquez-Fernandez E, Vos MR, Afanasyev P, Cebey L, Sevillano AM, Vidal E, Rosa I, Renault L, Ramos A, Peters PJ, Fernandez JJ, van Heel M, Young HS, Requena JR, Wille H (2016) The structural architecture of an infectious mammalian prion using electron cryomicroscopy. PLoS Pathog 12(9):ARTN

e1005835. https://doi.org/10.1371/journal.ppat.1005835

5. Dworkin S, Heath JK, deJong-Curtain TA, Hogan BM, Lieschke GJ, Malaterre J, Ramsay RG, Mantamadiotis T (2007) CREB activity modulates neural cell proliferation, midbrain-hindbrain organization and patterning in zebrafish. Dev Biol 307(1):127–141. https://doi.org/10.1016/j.ydbio.2007.04.026
6. Shabihkhani M, Lucey GM, Wei BW, Mareninov S, Lou JJ, Vinters HV, Singer EJ, Cloughesy TF, Yong WH (2014) The procurement, storage, and quality assurance of frozen blood and tissue biospecimens in pathology, biorepository, and biobank settings. Clin Biochem 47(4–5):258–266. https://doi.org/10.1016/j.clinbiochem.2014.01.002
7. Bayram B, Akdeniz SS, Diker N, Helvacioglu F, Erdem SR (2018) Effects of platelet-rich fibrin membrane on sciatic nerve regeneration. J Craniofac Surg 29(3):e239–e243. https://doi.org/10.1097/SCS.0000000000004256
8. Das G, Reuhl K, Zhou R (2013) The Golgi-Cox method. Methods Mol Biol (Clifton, NJ) 1018:313–321. https://doi.org/10.1007/978-1-62703-444-9_29
9. Karlsson O, Berg AL, Lindstrom AK, Hanrieder J, Arnerup G, Roman E, Bergquist J, Lindquist NG, Brittebo EB, Andersson M (2012) Neonatal exposure to the cyanobacterial toxin BMAA induces changes in protein expression and neurodegeneration in adult hippocampus. Toxicol Sci 130(2):391–404. https://doi.org/10.1093/toxsci/kfs241
10. Dworkin S, Malaterre J, Hollande F, Darcy PK, Ramsay RG, Mantamadiotis T (2009) cAMP response element binding protein is required for mouse neural progenitor cell survival and expansion. Stem cells (Dayton, Ohio) 27(6):1347–1357. https://doi.org/10.1002/stem.56
11. Malaterre J, Mantamadiotis T, Dworkin S, Lightowler S, Yang Q, Ransome MI, Turnley AM, Nichols NR, Emambokus NR, Frampton J, Ramsay RG (2008) c-Myb is required for neural progenitor cell proliferation and maintenance of the neural stem cell niche in adult brain. Stem cells (Dayton, Ohio) 26(1):173–181. https://doi.org/10.1634/stemcells.2007-0293

Chapter 14

Phenotypic Analysis of Early Neurogenesis in a Mouse Chimeric Embryo and Stem Cell-Based Neuruloid Model

Riley McMahon, V. Pragathi Masamsetti, and Patrick P. L. Tam

Abstract

Analyzing the impact of genetic mutations on early neurogenesis of mammalian embryos in conventional mouse mutant models is laborious and time-consuming. To overcome these constraints and to fast-track the phenotypic analysis, we developed a protocol that harnesses the amenability of engineering genetic modifications in embryonic stem cells from which mid-gestation mouse chimeras and in vitro neuruloids are generated. These stem cell-based chimera and neuruloid experimental models allow phenotyping at early developmental time points of neurogenesis.

Key words Embryos, Stem cells, Immunostaining, Confocal microscopy, Neurogenesis

1 Introduction

Conventional approaches to studying the consequence of gene loss- or gain-of-function on neural development require the generation of genetically modified mouse lines involving gene targeting of mouse embryonic stem cells (mESCs), germ line transmission of the mutant allele through the chimeras, and production by animal breeding [1]. These procedures are time-consuming and resource intensive. CRISPR–Cas9 genome editing performed directly in mouse zygotes, thereby bypassing the stem cell steps, has drastically reduced the time needed to create the desired genetic modifications and allows the use of the immediate generation of embryos and animals for experimentation. However, the utility of these models is confounded by the efficiency and the consistency of gene targeting among the embryos and F1 animals [2]. To circumvent these challenges, expeditious phenotypic analysis can be achieved using chimeric embryos generated from CRISPR–Cas9-edited mESCs [3].

Sebastian Dworkin (ed.), *Neurobiology: Methods and Protocols*, Methods in Molecular Biology, vol. 2746,
https://doi.org/10.1007/978-1-0716-3585-8_14,

Targeting a gene of interest using CRISPR–Cas9 enables rapid production of mono- and biallelic indels resulting in frameshift mutations that are frequently associated with loss of gene/protein function (*see* Sibbritt et al., [4] for the protocol to generate CRISPR–Cas9-edited mESCs). Using this gene editing technique, an individual locus or multiple gene loci can be modified in the genome of the mESCs. mESCs harboring the edited gene/s are incorporated into preimplantation embryos, followed by the transfer of chimeric embryos for further growth and development in utero, and harvested in the postimplantation period for the phenotypic analysis [5, 6]. Introducing mESCs to 8-cell stage embryos enhances the participation of the ESCs in the development of the inner cell mass/epiblast of the blastocyst, leading to high contribution of the ESC-derived cells to over 90% of the host tissues [7]. With the use of an appropriate host of which the cells are marked by a vital marker (e.g., fluorescent protein), the proportion of unmarked or differently marked ESC-derived tissues (i.e., the level of chimerism) can be quantified by imaging. Here, in this protocol, we used host 8-cell stage embryos that ubiquitously express a red fluorescent protein (RFP) named DsRed.t3 to distinguish between host and ESC-derived tissues and determine the extent of ESC contribution in the chimeric embryo. The morphological and molecular phenotype of chimeric embryos can be analyzed at a range of gestational ages/stages of postimplantation development, without resorting to using a conventional germ line mutant mouse model [5]. Using the chimeric embryo for analyzing the mutant phenotype overcomes the confounding factors of embryo lethality that limits the availability of experimental materials and facilitates the study of the functional impact of multiple/concurrent mutations in the same genome.

In conjunction with chimera analysis, in vitro stem cell-derived "neuruloids" are a valuable tool to investigate gene function in early neurogenesis [8]. Neuruloids are self-organized, in vitro three- dimensional structures generated by coculturing mESCs and wild-type extraembryonic endoderm (XEN) cells. These neuruloids recapitulate the interaction of extraembryonic endoderm tissue and the ESC-derived tissue to produce epiblast cells with anterior/neural characteristics that are competent for neuroectoderm differentiation [9]. The XEN cells in these organoids produce Wnt signaling antagonists to drive the differentiation of the ESC-derived cells to an anterior epiblast fate. The neuruloids can be produced at a large scale and are able to express early neural lineage markers (Fig. 1a) [10].

Like the chimeric embryo, the ESCs can be genetically modified to harbor loss/gain of gene function, either generically or conditionally time-/tissue-wise. Other genetic modifications such as inducible/enforced gene expression [11] or those enabling

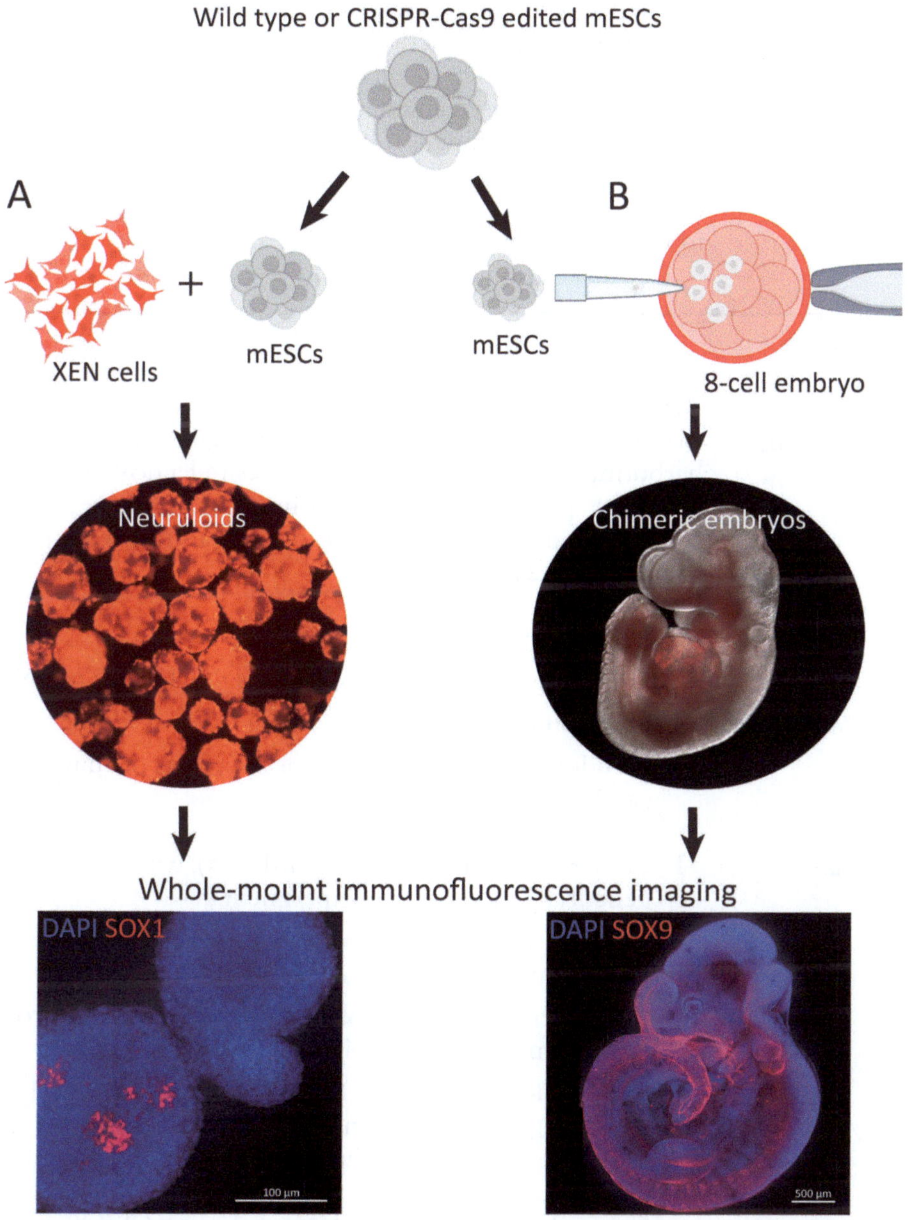

Fig. 1 Schematic depicting the use of mESCs to generate in vitro neuruloids and chimeric mid-gestation embryo. (**a**) XEN cells are cocultured with mESCs in 3D suspension culture, the DsRed.T3 expressing XEN cells colonize the superficial tissues of the neuruloids and secrete Wnt antagonists. (**b**) Chimeric mouse embryos are generated through the injection of mESCs into 8-cell preimplantation embryos expressing DsRed.T3 and then transferred into the reproductive tract of recipient female mice for further development. The chimeras are then collected at a range of embryonic stage (e.g., E9.5 as shown) for analysis. Both neuruloids and chimeras can then be processed for whole-mount immunofluorescence visualization of the expression of neural lineage markers (e.g., SOX1 or SOX9) and other multi-omics analyses. mESCs, mouse embryonic stem cells; XEN, extraembryonic endoderm cells (Figure was created with BioRender.com)

assays for protein–DNA interaction by DamID-seq [12] and protein–protein interactions by BioID [13] can be applied to ESCs to study function of the gene and its interactions. Overall, these stem cell-based chimeras and neuruloid models are amenable for validating the function of genes and gene regulatory networks of interest and are effective experimental tools to investigate the cellular and molecular mechanism of early neurogenesis in vivo and in vitro, respectively.

2 Materials

2.1 Choice of Embryonic Stem Cells

Begin with mESCs that harbor the requisite genetic modifications that fit the purpose of the experimental studies, such as those harboring CRISPR–Cas9-edited genes [4] a doxycycline inducible transgene [11] for overexpression analysis or other modification suitable for downstream analysis of gene and network function (*see* **Notes 1** and **2**).

2.2 Mice

1. ARC/s (outbred) female mice.
2. Homozygous DsRed.T3 expressing male mice (*see* **Note 3**).
3. Vasectomized male mice.
4. Pseudopregnant female mice as recipients of chimeric embryos (*see* **Note 4**).

2.3 Cell Culture

1. Mouse embryonic stem cells (mESCs): wild type as control and genetically modified for experimental.
2. Extraembryonic endoderm stem cells (XEN cells) derived from E3.5 mouse blastocysts, as described [14].
3. Mouse embryonic fibroblasts (MEFs) derived from E13.5 mouse embryos, as described [15].
4. mESC culture medium: DMEM (Dulbecco's Modified Eagle Medium), 12.5% heat-inactivated fetal bovine serum (FBS), 0.1 mM β-mercaptoethanol, 1x nonessential amino acids (NEAA), 1x nucleosides, and 1000 U/mL leukemia inhibitory factor (LIF). 2i media inhibits selective GSK3β and MEK1/2 and is prepared by addition of 3 μM CHIR99021 and 1 μM PD0325901 to mouse mESC LIF containing media.
5. XEN culture medium: Advanced RPMI 1640 (Roswell Park Memorial Institute Medium), supplemented with 20% fetal bovine serum, 1 mM L-glutamine, 0.1 mM β-mercaptoethanol, 1 mM sodium pyruvate; 1% (vol/vol) penicillin–streptomycin. For derivation of XEN cell lines, 24 ng/mL FGF4 and 1 μg/mL heparin were added.

6. N2B27 medium: DMEM-F12, 0.5× N2 supplement, 0.5× B27 supplement, 0.5 mM L-glutamine, 1% (vol/vol) penicillin–streptomycin, 0.5× NEAA, 0.1 mM β-mercaptoethanol.
7. 15 mL Falcon tubes.
8. 0.1% gelatin (cell culture grade).
9. AggreWell™ 400 24 wells.
10. TrypLE™ Select.
11. Dulbecco's phosphate-buffered saline (DPBS).
12. 10 cm tissue culture plates (adherent and low adherence).
13. Humidified incubator (37 °C, 5% CO_2).
14. Orbital shaking platform in incubator.
15. Hemocytometer or automatic cell counting machine.

2.4 Mice Dissection and Microinjection

1. Mouse dissection equipment: iridectomy scissors, dissection scissors, fine forceps, 32G blunt end needle, blunt 26G needle, 1 mL syringe, 5 mL syringe.
2. 60 mm tissue culture dishes.
3. M2 media.
4. M16 media.
5. PB1 media, *see* [16].
6. Mineral oil (embryo grade).
7. Dissecting microscope.
8. Pulled glass mouth pipettes, *see* [4].
9. Eppendorf TransferMan NK 2 system with an Olympus IX70 inverted microscope.
10. VacuTip I tips (for holding embryos).
11. TransferTips-ES (for injecting mESCs).
12. Ketamine, xylazil (100 mg/kg body mass), buprenorphine (0.05–0.1 mg/kg body mass).
13. Heat pad, medical stitches, 30G needles, 27G needles, forceps, clamps, 70% ethanol, normal saline.

2.5 Immuno-fluorescence Imaging

1. 4% paraformaldehyde (PFA) in PBS.
2. PBST: DPBS with 0.2% Tween 20.
3. Methanol.
4. Bleaching solution: one part 30% hydrogen peroxide to two parts methanol.
5. Dent's fixative: one part dimethyl sulfoxide (DMSO) to four parts methanol.

6. Blocking solution: 0.2% bovine serum albumin (BSA), 20% DMSO, 0.4% Triton X-100 in DPBS.
7. Primary antibodies:
   - Rabbit anti-TFAP2A (1:1000 in PBST).
   - Mouse anti-neurofilament (NF-M) (1:500 in PBST).
8. Secondary antibodies:
   - Alexa Fluor 633 donkey anti-rabbit IgG (1:1000 in PBST).
   - Alexa Fluor 488 donkey anti-mouse IgG (1:1000 in PBST).
9. DAPI: 4′,6-diamidino-2-phenylindole (1:1000 in PBST).
10. BABB: one part benzyl alcohol to two parts benzyl benzoate.
11. SteREO Lumar.V12 microscope.
12. Zeiss Cell Observer Spinning Disc Confocal Microscope.
13. Ibidi chamber wells.

3 Methods

3.1 Generating Neuruloids

1. Maintain mESCs in 2i/LIF medium on gelatin in 10 cm tissue culture plates for at least two passages prior the generation of neuruloids (*see* **Note 5**).
2. Maintain XEN cells on gelatin in XEN culture medium in 10 cm tissue culture plates.
3. To generate the neuruloids, wash mESCs and XEN cells with 10 mL of DPBS for 1 min, before adding 2 mL of TrypLE select to each plate and incubating for 5 min at 37 °C.
4. Use serum containing media to inactivate the TrypLE select, then add dissociated cells to a 15 mL tube, and centrifuge at 100 g for 5 min.
5. During this time, add 1 mL of N2B27 media to each AggreWell™ 400 well (24-well plate) and centrifuge at 1000 g for 3 min to remove bubbles.
6. Resuspend cells in DPBS and count cell number using hemocytometer or automatic cell counter.
7. Add approximately 2.5×10^6 mESCs and 0.5×10^6 XEN cells per AggreWell™ (*see* **Note 6**) to a 15 mL Falcon tube and spin down for 5 min at 100 g.
8. Resuspend the cell mixture in 1 mL of N2B27 media per AggreWell™ and pipette 1 mL into each AggreWell™ 400 well. Centrifuge the plate at 100 g for 3 min to bring cells to the bottom. Put the plate in the incubator at 37 °C.
9. After 48 h of culture, transfer the cell aggregates from the AggreWell™ to low adhesion 10 cm plates using a 1000 μL pipette tip with a small cut at the top of the tip (*see* **Note 7**). Fill

each plate to 10 mL of N2B27 medium, add 3 μM of CHIR99021, and incubate on a shaking platform (50–60 rpm).

10. After 24 h, transfer neuruloids and medium to 15 mL tube and stand for 10 min to allow the organoids to settle in the bottom before aspirating off the spent media and replacing with 10 mL of fresh N2B27 medium (without CHIR99021) (*see* **Note 8**). Return the neuruloids to shaking incubator (50–60 rpm) for further culture.
11. Collect neuruloids after at 96–144 h (total culture time; *see* **Note 9**), stand them in a 15 mL Falcon tube until they have settled to the bottom, and wash three times with DPBS.
12. For molecular analysis (that requires RNA/DNA extraction), snap freeze cell clumps in 1.5 mL Eppendorf tube by immersing in liquid nitrogen. For immunofluorescence imaging, proceed to Subheading 3.4 for processing the materials.

3.2 Generating Chimeras

1. Set up time-mating of ARC/s (outbred) female mice and homozygous DsRed.T3 male mice. Collect RFP-expressing 8-cell embryos from female mice at 2.5 days post-coitus.
2. Sacrifice pregnant female mice and excise the uterine horn and the oviduct between the cervical junction and the oviductal infundibulum, place the explant in a 60 mm dish with M2 media, remove the fat, and bisect the explant at the uterine side of the oviductal–uterine junction.
3. Flush the uteri from both ends of the uterine horn with M2 media using blunt ended 26G hypodermic needles and flush the oviducts with M2 medium using a blunt 32G needle inserted into the infundibulum opening. Collect embryos using a month-controlled glass micropipette.
4. Place 8-cell embryos in drops of M16 medium under mineral oil in 60 mm dish and incubate at 37 °C in 5% CO_2 in air until ready for injection of ESCs on the same day.
5. Dissociate mESCs grown in 2i/LIF media and resuspend 2×10^6 cells in ~1 mL media, and then place tube on ice (to remove the embryonic fibroblast feeder cells, *see* **Note 10**).
6. Place 8–10 embryos and mESCs in small drops of M16 media underneath mineral oil using a mouth-controlled micropipette. Using an Eppendorf TransferMan NK 2 system, hold embryos using VacuTip I tips and inject 10–15 mESCs using TransferTips-ES into the space between the blastomeres (*see* **Note 11**). Culture injected embryos in drops of KSOM or M16 media underneath mineral oil overnight at 37 °C, 5% CO_2 (*see* **Note 12**).

7. On the following day, anesthetize pseudopregnant female mice with ketamine/xylazil (100 mg/kg body mass) and buprenorphine (for analgesic, 0.05–0.1 mg/kg body mass).
8. While the animal is sedated, make an incision through the body wall to retrieve the uterine horn, clamp the fat tissue to stabilize the exteriorized horn, and move the mouse to underneath a dissecting microscope.
9. Pierce the wall of the uterus using a 27G needle and transfer embryos in M2 media inside the uterus through the wound using a glass mouth-controlled micropipette (*see* **Notes 13** and **14**).
10. Return the uterus to the peritoneal cavity of the recipient mouse and close the body wall with sutures and wound clips, inject 500 μL of normal saline subcutaneously, and monitor the postsurgery recovery.
11. Dissect the uterus to harvest the chimeric embryos in PB1 medium on the gestational day of the experimental end point (e.g., E9.5, Fig. 1b) and processed as appropriate for the specific phenotype analysis (*see* **Note 15**).

3.3 Analyzing mESC Contribution to Chimeric Embryos

1. Image the chimeric embryos in a 60 mm plate in DPBS using SteREO Lumar.V12 microscope for brightfield and DsRed.T3 expression (*see* **Note 16**). Make sure to use the same fluorescence exposure settings for each embryo in the same batch of experiment.
2. Import the image file into ImageJ (download at https://imagej.nih.gov/ij/download.html), outline the embryo using freehand ROI tool, and measure area, mean fluorescence, and intensity density (IntDen) of the fluorescent DsRed.T3 channel using Analyze > Measure (*see* Fig. 2).
3. Outline a small area of the image that has no fluorescence; use this as a background value (do this multiple times to get an average value).
4. Calculate corrected total cell fluorescence (CTCF) = integrated density − (area of selected cells × mean fluorescence of background readings).
5. Use subtracted value to compare fluorescence intensity between chimeric embryos at the same stage.

3.4 Protein Expression by Immunofluorescence and Visualization by Imaging

1. Fix the embryos or neuruloids in 4% PFA for at least 2 h at 4 °C.
2. Wash the samples in cold PBST three times, and then, transfer through a series of higher methanol concentrations in PBST (25%, 50%, 75%) before leaving in 100% methanol overnight at 4 °C.

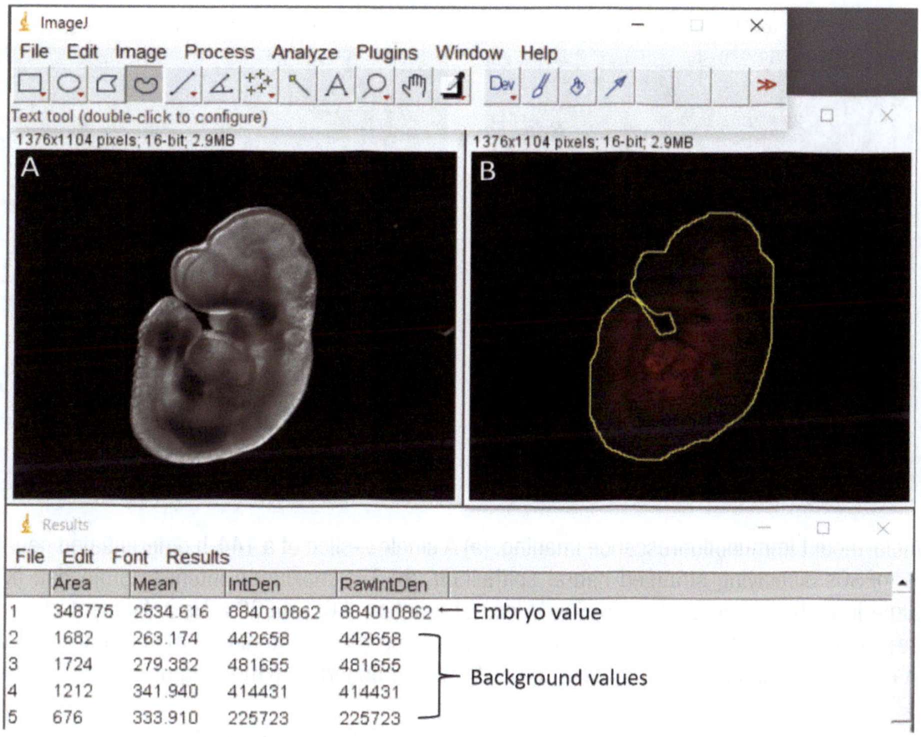

Fig. 2 ImageJ fluorescence analysis of chimeric embryos. (**a**) Brightfield channel and (**b**) red fluorescence channel. The DsRed.T3 expression of an individual E9.5 chimeric embryo is quantified by measuring the intensity density of the embryos and subtracting the background fluorescence. The intensity of red fluorescence is negatively correlated with the level of contribution by the nonfluorescent ESCs

3. Transfer samples to bleaching solution for 24 h at 4 °C.
4. Wash samples three times with cold 100% methanol for 10 min, and then, add Dent's fixative for 24 h at 4 °C.
5. Block samples for 1 h in blocking solution at room temperature.
6. Add primary antibodies such as rabbit anti-TFAP2A and mouse anti-neurofilament to blocking solution without Triton (*see* **Note 17**).
7. Incubate embryos with primary antibodies. For small embryos and neuruloids, incubate overnight; for larger embryos, incubate for up to 4 days.
8. Wash in PBST for 3 h, changing buffer every 30 min.
9. Add secondary antibodies such as Alexa Fluor 633 donkey anti-rabbit IgG (1:1000 in PBST), Alexa Fluor 488 donkey anti-mouse IgG (1:1000 in PBST), as well as DAPI (for DNA staining, 1:1000 in PBST) in blocking solution without Triton overnight at room temperature (*see* **Note 18**).
10. Wash samples in PBST for 3 h, changing buffer every 30 min.

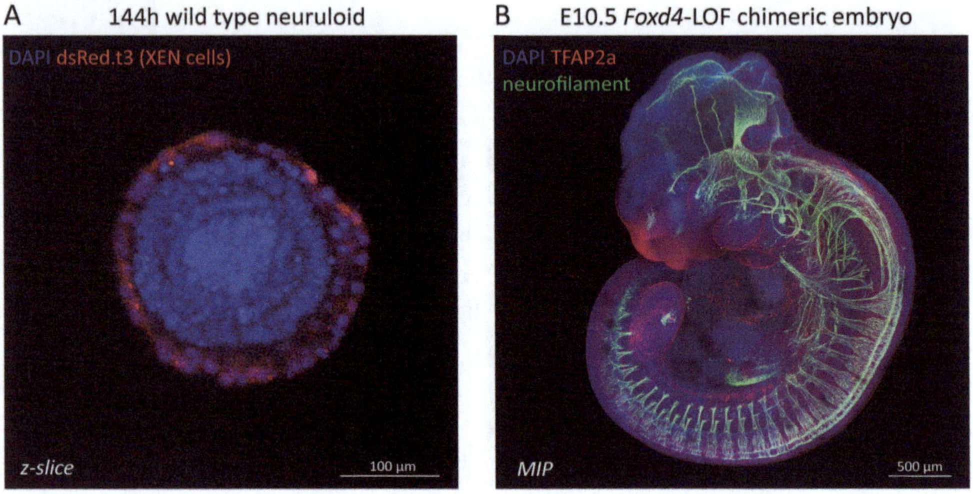

Fig. 3 Whole-mount immunofluorescence imaging. (**a**) A single z-slice of a 144-h differentiated neuruloid with wild-type mESCs displaying stratified neural epithelium. (**b**) The maximum intensity projection (MIP) of an E10.5 chimeric embryo generated from *Foxd4* loss of function (LOF) mESCs. Neurofilament staining shows normal segmental spinal nerves, while the closure defects in the rostral neural tube are accompanied by a reduction of TFAP2a-expressing neural crest derivatives and abnormal patterning of the cranial nerves

11. Dehydrate samples in 50% and then 75% methanol in PBST, followed by 100% methanol.
12. Clear samples in 50% BABB in methanol for 10 min and then 100% BABB overnight.
13. Image the neuruloids or chimeric embryos using a confocal microscope in chamber wells: For neuruloids and small embryos (E6.5–E8.25), use 20× magnification using z-stack slices and tiling; for larger chimeras (E8.5–E11.5), use 10× magnification.
14. Use imaging software to reconstruct 3D images using z-stack and tiling to visualize embryonic structures such as neural tube, craniofacial primordia, neural crest derivatives, etc.
15. Protein localization in neuruloids and chimeric embryos can be visualized by overlaying the immunofluorescence signal over the embryonic structures (Fig. 3a, b).

4 Notes

1. CRISPR–Cas9 editing using single guide RNAs can generate small insertions and deletions efficiently at the desired locus. To generate larger deletions such as whole exon deletions, dual-guide RNAs can be introduced into mESCs using a single plasmid that can express both guide RNAs [17].

2. Due to the possibility of off-target effects of CRISPR–Cas9 editing, it is advisable to generate mESC clones usings multiple different gRNAs targeting the same gene [10].
3. Homozygous DsRed.T3 expressing male mice are essential to generate whole liters of host embryos expressing the red fluorescent protein (RFP). Introducing unmarked or differently marked mESCs into the RFP-positive host embryo allows selection of high ESC contribution chimeras for phenotypic analysis.
4. Vasectomized male mice should be time-mated with female mice to generate the pseudopregnant recipients of chimeric embryos 1 day after the mating of the female mice that provide the recipient embryos. This is to enhance implantation efficiency.
5. Culturing the mESCS in 2i/LIF media for two passages is essential so the cells are maintained at ground state pluripotency.
6. Exact numbers of mESCs and XEN cells to add to each AggreWell™ are to be optimized for different cell lines with consideration of the proliferation rate of the cell lines chosen for experimentation.
7. XEN cells will colonize the superficial cell population of the neuruloids. XEN cell line derived from mice embryos harboring a fluorescent marker (such as DsRed.T3) enables verification of the localization of the XEN cells in live neuruloids.
8. Make sure to not leave the neuruloids to settle for longer than 10 min when changing media so that single cells in suspension do not settle with the neuruloids.
9. Leaving the neuruloids to culture longer than 96 h leads to further regionalization and neural epithelial tissue to develop (Fig. 3a) [9]. The media needs changing for the extended culture.
10. If the mESCs are cultured on mouse embryonic fibroblasts (MEFs), the MEFs can be removed from the culture by plating ESC suspension on gelatin for 15 minutes and collecting from the supernatant the ESCs for the injection into host embryos.
11. The number of mESCs injected into each 8-cell embryo is optimized for the requisite level of mESC contribution in different experiments. Generally, for high level of mESC contribution in E10.5 embryo, 15 mESCs per embryo may be injected. It is also possible to inject mESCs into blastocyst stage, though the level of contribution would generally be lower.

12. Try to limit the time that mouse embryos are outside of the incubator by handling embryos in small batches of 8–10 for microinjection. With increased experience of the operator, more embryos can be injected in each batch.
13. The micropipette used for transferring embryos to the uterus should be wide enough for transferring multiple embryos that are lined up in the tip region in the smallest volume of medium, and the tip of the micropipette should be flame polished to facilitate navigating the micropipette through the needle wound in the uterine wall.
14. Transfer no more than 12 embryos per uterine horn to achieve implantation of about six to eight embryos per transfer to avoid overcrowding which may impede embryo growth.
15. For the chimeric embryo, while the developmental age of preimplantation host embryo is ahead by about 1 day of the gestational age of the pseudopregnant female mice, the development of chimeric embryo will be synchronized with the gestational age in postimplantation period. Embryo collection should be timed by the gestational age of the recipient mice.
16. When analyzing the contribution of mESCs to chimeric embryos, check for any bias in the contribution of the mESCs to the three germ layer derivatives. The lack of or reduced contribution to certain tissue may be attributed to altered lineage propensity, rather than aberration of tissue differentiation.
17. Choose the concentration of each individual primary antibody based on the manufacturer's instructions for immunofluorescence imaging. The appropriate concentration may need to be optimized based on the age of the embryo. When selecting other antibodies, be sure to use a different host species for each primary antibody.
18. When selecting secondary antibodies for immunofluorescence imaging, it is important to not select fluorophores with overlapping excitation spectra as your chimeric embryo host cells (e.g., we would not use Alexa Fluor 594 donkey anti-rabbit IgG as our host cells express DsRed.T3 with an excitation wavelength of 560 nm). Refer to a fluorochrome chart when selecting your secondary antibodies.

References

1. Hall B, Limaye A, Kulkarni AB (2009) Overview: generation of gene knockout mice. Curr Protoc Cell Biol Chapter 19:Unit 19.12 19.12.1–17
2. Papathanasiou S, Markoulaki S, Blaine LJ et al (2021) Whole chromosome loss and genomic instability in mouse embryos after CRISPR-Cas9 genome editing. Nat Commun 12(1): 5855. https://doi.org/10.1038/s41467-021-26097-y
3. Oji A, Noda T, Fujihara Y et al (2016) CRISPR/Cas9 mediated genome editing in

ES cells and its application for chimeric analysis in mice. Sci Rep 6(1):31666. https://doi.org/10.1038/srep31666
4. Sibbritt T, Osteil P, Fan X et al (2019) Gene editing of mouse embryonic and epiblast stem cells. In: Mouse cell culture. Springer, Cham, pp 77–95
5. Tam PP, Rossant J (2003) Mouse embryonic chimeras: tools for studying mammalian development. Development 130(25):6155–6163
6. Eckardt S, McLaughlin KJ, Willenbring H (2011) Mouse chimeras as a system to investigate development, cell and tissue function, disease mechanisms and organ regeneration. Cell Cycle 10(13):2091–2099
7. Poueymirou WT, Auerbach W, Frendewey D et al (2007) F0 generation mice fully derived from gene-targeted embryonic stem cells allowing immediate phenotypic analyses. Nat Biotechnol 25(1):91–99. https://doi.org/10.1038/nbt1263
8. Rossant J, Tam PP (2021) Opportunities and challenges with stem cell-based embryo models. Stem Cell Rep 16(5):1031–1038
9. Bérenger-Currias NM, Mircea M, Adegeest E et al (2022) A gastruloid model of the interaction between embryonic and extra-embryonic cell types. J Tissue Eng 13: 20417314221103042
10. McMahon R, Sibbritt T, Aryamanesh N et al (2021) Loss of Foxd4 impacts neurulation and cranial neural crest specification during early head development. Front Cell Dev Biol 9: 777652
11. Iacovino M, Bosnakovski D, Fey H et al (2011) Inducible cassette exchange: a rapid and efficient system enabling conditional gene expression in embryonic stem and primary cells. Stem Cells 29(10):1580–1588
12. Greil F, Moorman C, van Steensel B (2006) DamID: mapping of in vivo protein–genome interactions using tethered DNA adenine methyltransferase. Methods Enzymol 410: 342–359
13. Kim DI, Jensen SC, Noble KA et al (2016) An improved smaller biotin ligase for BioID proximity labeling. Mol Biol Cell 27(8):1188–1196
14. Niakan KK, Schrode N, Cho LT et al (2013) Derivation of extraembryonic endoderm stem (XEN) cells from mouse embryos and embryonic stem cells. Nat Protoc 8(6):1028–1041
15. Conner DA (2001) Mouse embryo fibroblast (MEF) feeder cell preparation. Curr Protoc Mol Biol 51(1):23.2.1–23.2.7
16. Masamsetti VP, Tam PP (2022) Identification and visualization of protein expression in whole mouse embryos by immunofluorescence. In: Epiblast stem cells. Springer, New York, pp 39–45
17. Adikusuma F, Pfitzner C, Thomas PQ (2017) Versatile single-step-assembly CRISPR/Cas9 vectors for dual gRNA expression. PLoS One 12(12):e0187236

Chapter 15

Active Induction of a Multiple Sclerosis-Like Disease in Common Laboratory Mouse Strains

Jacqueline M. Orian, Dain L. Maxwell, and Vernise J. T. Lim

Abstract

Experimental autoimmune encephalomyelitis (EAE) is a neuroinflammatory disease with facets in common with multiple sclerosis (MS). It is induced in susceptible mammalian species, with rodents as the preferred hosts, and has been used for decades as a model to investigate the immunopathogenesis of MS as well as for preclinical evaluation of candidate MS therapeutics. Most commonly, EAE is generated by active immunization with central nervous system (CNS) antigens, such as whole CNS homogenate, myelin proteins, or peptides derived from these proteins. However, EAE actually represents a spectrum of diseases in which specific combinations of host/CNS antigen exhibit defined clinical profiles, each associated with unique immunological and pathological features. Similar to MS, EAE is a complex disease where development and progression are also modulated by environmental factors; therefore, the establishment of any given EAE variant can be challenging and requires careful optimization. Here, we describe protocols for three EAE variants, successfully generated in our laboratory, and provide additional information as to how to maintain their unique features and reproducibility.

Key words Experimental autoimmune encephalomyelitis (EAE), Multiple sclerosis (MS), C57Bl/6, NOD/ShiLt, BALB/c, Active immunization, Encephalitogenic peptide, Murine EAE, Neuroinflammation

1 Introduction

EAE is a CNS autoimmune disease generated in experimental animals, which has features in common with the human disorder MS [1]. It was originally designed [2] to investigate the paralytic inflammatory demyelinating disease occasionally observed during the development of a rabies vaccine but since its original inception has been constantly modified and improved to act as a model of the immune-mediated CNS demyelination associated with MS [3–6].

The original version of the chapter has been revised. A correction to this chapter can be found at https://doi.org/10.1007/978-1-0716-3585-8_19

Sebastian Dworkin (ed.), *Neurobiology: Methods and Protocols*, Methods in Molecular Biology, vol. 2746,
https://doi.org/10.1007/978-1-0716-3585-8_15,

Consequently, EAE has now become the most widely accepted MS animal model. The disease is generated by sensitization with CNS antigens, including spinal cord homogenate, purified myelin proteins, or their immunodominant epitopes, in susceptible species such as mice, rats, guinea pigs, hamsters, marmosets, and macaques [3–9]. It exhibits symptoms in common with MS, including ambulatory difficulties, impaired balance, as well as bladder and bowel dysfunction. Pre-onset neuropsychological deficits have also recently been reported using experimental paradigms of anxiety-like behavior [10–15], which concords with the novel view that depression is a primary disease manifestation in MS [16, 17].

The pathological hallmarks of EAE are also reminiscent of those of MS [18]. EAE is characterized by meningeal and perivenous inflammation. In the white matter, this results in the development of severe demyelinating lesions dominated by T cells (especially $CD8^+$), B cells, and macrophages containing myelin debris. These hallmarks are associated with prominent and widespread microglial and astrocytic reactivity, as well as axonal damage and truncation, extending into the normal appearing white matter. In the gray matter, by contrast, lesions remain small or never develop, but extensive microglial and astrocytic reactivity is observed, associated with demyelination, axonal injury, and neuronal loss, beginning early in disease development [19–21].

EAE is a complex disease and, similarly to MS, is incompletely understood [22, 23]. It is induced most commonly by active immunization, namely, by subcutaneous injection of myelin proteins or peptides in adjuvant. Alternatively, it can be generated by adoptive transfer which consists of intravenous delivery into naïve recipients, of activated $CD4^+$ or $CD8^+$ cells [24–26] originating from draining lymph nodes of actively induced mice, and then restimulated with their cognate antigens. EAE never manifests spontaneously, except in a restricted number of genetically modified murine models where mechanisms that normally suppress tissue-specific autoreactivity are bypassed [26–28]. Rodents remain the preferred host but display a wide spectrum of variants differing in terms of degrees of disease susceptibility and clinical profiles [29–31], which are determined by both genetic and nongenetic elements [22, 32]. EAE variants (Fig. 1) are characterized by the following:

1. *Host—encephalitogenic protein/peptide specificity*: Each mouse/rat strain exhibits sensitivity to distinct antigenic protein/peptides [33, 34]. For example, while the C57BL/6, Biozzi ABH, NOD/Lt, SJL/J mice, and DA (dark agouti) rats are sensitive to sequences derived from myelin oligodendrocyte glycoprotein (MOG [35]; a quantitatively minor myelin component), each host responds to a unique MOG sequence, namely, MOG_{35-55} for C57BL/6 and NOD/Lt mice, but MOG_{92-106} for the SJL/J mouse, or MOG_{8-22} and MOG_{1-125} ("whole" MOG) for Biozzi ABH mice or DA rats

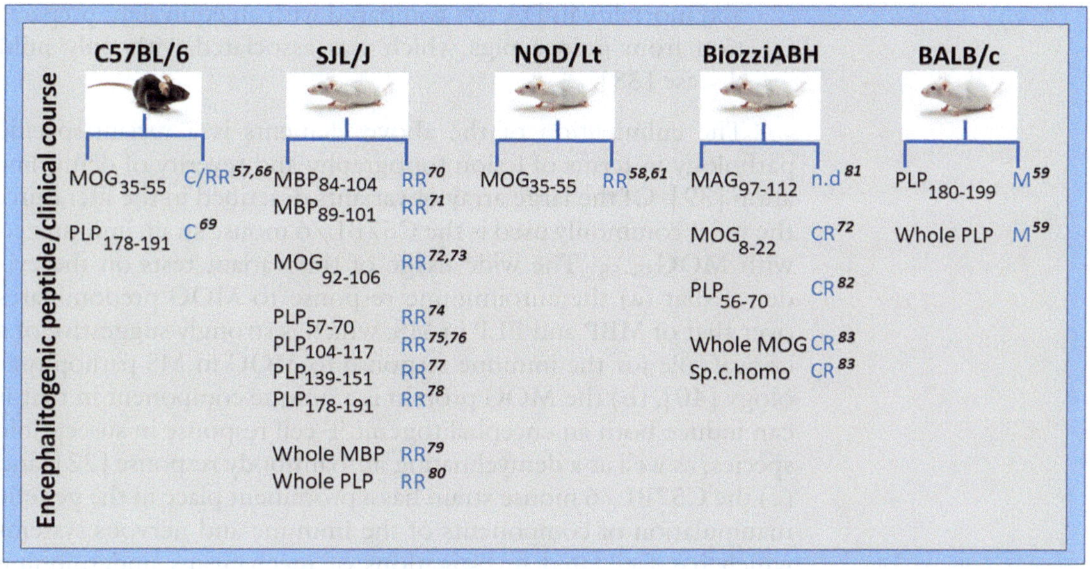

Fig. 1 EAE exists as a number of variants generated by altering combinations of strain and encephalitogenic peptide. Shown is a list of murine EAE variants generated in common mouse strains. Note that (**a**) each mouse strain responds to a different range of encephalitogenic peptides (or whole proteins), (**b**) any given encephalitogenic species is associated with a distinct clinical course in each mouse strain, and (**c**) the SJL/J mouse strain exhibits relatively higher promiscuity. Not shown is that the source of encephalitogenic species also modulates the disease profile. The list is for the purposes of illustration and not exhaustive and multiple additional variants have been reported in the literature. Abbreviations: C = chronic (acute phase followed by permanent disability); CR = chronic relapsing (acute phase followed by permanent disability punctuated by intermittent exacerbations); M = monophasic (acute disease followed by significant/complete recovery without subsequent attacks); RR = relapsing–remitting (acute disease followed by clearly defined exacerbations interrupted by periods of significant/complete recovery); n.d. = not determined; sp.c.homo = spinal cord homogenate; MAG = myelin-associated glycoprotein. Source: Ref. [26, 29, 62] and additional manuscripts [57–59, 61, 66–83] shown in the figure

[29–31]. However, this specificity can be relatively broad in certain strains as in SJL/J mice where EAE can be induced with a wide range of peptides derived from multiple myelin proteins [29–31], other than MOG and including the quantitatively major myelin components proteolipid protein (PLP [36]) and myelin basic protein (MBP [37]).

2. *Host—encephalitogenic protein/peptide and clinical profile specificity.* Secondly, each host and encephalitogenic protein/peptide combination results in a defined clinical profile, for example, chronic progressive in the case of the C57BL/6 × $MOG_{35–55}$ and DA rat × $MOG_{1–125}$ combinations or chronic-relapsing for the NOD/Lt × $MOG_{35–55}$, SJL/J × $MOG_{92–106}$, or Biozzi ABH × $MOG_{8–22}$ and whole MOG combinations [29–31].
3. *Source of antigen:* Finally, it was demonstrated that the source of antigen also modulates the disease, when spinal cord homogenate from rat resulted in higher incidence, severity,

and mortality in DA rats compared with an equivalent preparation from guinea pigs which was associated with only mild disease [38].

The culmination of the above elements is a variant-specific pathology in terms of lesion topography and severity of demyelination [39]. Of the large array of variants described in the literature, the most commonly used is the C57BL/6 mouse strain immunized with MOG_{35-55}. The wide usage of this variant rests on the evidence that (a) the autoimmune response to MOG predominates over that of MBP and PLP in MS, which is strongly suggestive of a critical role for the immune response to MOG in MS pathophysiology [40], (b) the MOG protein is a unique component in that it can induce both an encephalitogenic T-cell response in susceptible species, as well as a demyelinating autoantibody response [22], and (c) the C57BL/6 mouse strain has a prominent place in the genetic manipulation of components of the immune and nervous systems which has facilitated investigations of mechanisms underpinning the pathogenesis of chronic inflammation and tissue damage [41, 42]. It is also important to remember that disease development is influenced by the environment, including the gut microbiome [43]. Therefore, differences in environments such as those that occur between conventional, specific pathogen free (SPF), or germ-free animal facilities will impact disease profiles, most noticeably in terms of timing of clinical onset, degree of severity, or disease duration.

The sensitization to myelin antigens requires the use of strong adjuvants. These are agents which, in combination, strongly potentiate an immune reaction by augmenting the immunogenicity of a given antigen [44, 45]. A major function of adjuvants is to act as antigen depots, whereby the slow release from the injection site ensures persistent stimulation of the immune system. They are known to have additional roles, namely, in enhancing antigen-presenting cell function by upregulation of MHC and co-stimulatory molecule expression, as well as increasing cytokine expression by macrophages, including IL-1 and IL-12, which promote cell mediated immune responses [44]. In EAE, antigens are delivered in complete Freund's adjuvant (CFA), which combines the above properties. CFA is prepared from incomplete Freund's adjuvant (IFA; 85% paraffin oil and 15% mannide monooleate as an emulsifier), supplemented with *Mycobacterium tuberculosis* (*M. tuberculosis*) [46]. This bacterium is heat inactivated prior to being added at a concentration of 1 mg/mL. *M. tuberculosis* potentiates the adjuvanticity of CFA via the Th1 response to myelin antigens and by additional effects as a strong inducer of IL-12. However, it is likely that this component also contributes to blood–brain barrier (BBB) permeability [44]. Finally, EAE usually requires the administration of the toxic protein from the bacterium

Bordetella pertussis (*B. pertussis*), which is a virulence factor for the bacterium that causes whooping cough [47]. Pertussis toxin (PTx) has complex effects, the predominant one in the context of EAE being an increase in BBB permeability. It is believed to also stimulate the Th1 response, via IL-12 induction [44]. Although the generation of most EAE variants requires PTx, certain SJL/J x PLP variants [30] (but not MBP x SJL/J) are notable exceptions, where chronic-relapsing disease can be induced in the absence of this adjuvant.

Protocols for active EAE induction generally utilize female mice only, to reproduce the strong female bias in susceptibility to MS [1]. The immunization schedule consists of an injection with antigenic peptide emulsified in CFA, followed by PTx delivery in saline solution via a separate injection. Disease development consists of a sequence of generally well-identified events [30], beginning with an induction phase where priming of myelin epitope-specific $CD4^+$ T cells occurs. This is followed by an effector phase, which sees the infiltration of activated myelin-specific T cells into the CNS by their extravasation across the BBB. This provokes the generation of a highly pro-inflammatory environment by the production of myelin-specific T cell-derived cytokines and chemokines, resulting in an influx of peripheral mononuclear phagocytes into the CNS parenchyma, followed by activation of peripheral monocytes/macrophages together with CNS-resident microglial cells. The combined effect of the cytotoxic activity of cytokines from T cells and monocytes and phagocytic activity of activated mononuclear cells is the demyelination of CNS axonal tracts.

Although EAE and MS have many common features, it is important to remember that they are different diseases [48–51]. While EAE is regarded as recapitulating neuroinflammation, MS is a far more complex disease underpinned by multiple factors, including immunological, neurodegenerative, vascular, and environmental components and where the pathophysiology is still incompletely understood [18, 52, 53]. The model has attracted its fair share of criticism because, (a) as mentioned, EAE is not spontaneous and requires the use of strong immune adjuvants in order to develop; (b) EAE only partially recapitulates MS and different variants of the model are required to investigate different clinical, immunological, and pathological MS facets; (c) the model makes use of inbred strains to maintain reproducibility of clinical profile and immune response characteristic of each variant, thereby losing the genetic heterogeneity inherent in MS populations; and (d) importantly, the model has not been reliably predictive of the efficacy of candidate MS therapeutics [4, 22, 23, 54]. On the other hand, there is no doubt that EAE, particularly with the use of genetically modified mouse lines, has provided valuable insight into mechanisms of immune-mediated injury, including mechanisms of BBB loss of function, differentiation between $CD4^+$ and $CD8^+$ subsets, and the role of the Th17 subset and associated cytokine/chemokine networks involved in their induction and downstream signalling [55, 56].

Given the complexity of MS and the limited understanding of its immunopathogenesis, it is unlikely that a more accurate model of the disease can be generated at this point. The current view of EAE is that it needs to be used rationally, by recognizing its limitations and incorporating these limitations into experimental designs. Questions that must be considered include (a) which aspect of the disease is being addressed and which EAE variant has the potential to recapitulate the particular disease aspect/stage or (b) what are the specific immune defects inherent in the various mouse strains and how can a given genetic background impact on the immune question of interest? With this perspective in mind, EAE can prove to be a useful tool to generate proof of concept for specific mechanisms underlying MS pathophysiology and explore repair and neuroprotective strategies [1, 22, 28].

Here, we describe the generation of active EAE in commonly used variants, namely, C57BL/6 and NOD/Lt mice with $MOG_{35–55}$ and BALB/c mice with $PLP_{180–199}$.

2 Materials

2.1 Mouse Strains

Female C57BL/6J, NOD/ShiLtJ or BALB/cJ mice (depending on the EAE variant to be generated), aged 9–10 weeks and weighing a minimum of 21 g at baseline (*see* **Note 1**).

Because of the influence of environment on disease expression, it is important to use mice that are routinely verified for genetic authenticity. Protocols described below have been optimized using mouse strains sourced from the Jackson Laboratory, hence the suffix "J" appended to each strain name.

2.2 Stock Reagents

The preparation of stock reagents is performed at room temperature (unless otherwise indicated) on a clean working surface. All lyophilized components should be reconstituted using deionized water. Gloves should be worn. Stock reagents remain stable for extended periods when stored appropriately.

1. Spatula, cleaned and dried.
2. Sterile cryogenic tubes of 2 mL capacity.
3. *Heat inactivated, lyophilized M. tuberculosis, strain H37Ra*: This reagent is supplied in ampules, each containing 100 mg. Store at 4 °C and refer to manufacturer's instructions for shelf life.

4. *Complete Freund's adjuvant*: CFA is usually supplied in vials each containing 10 mL and 1 mg/mL of *M. tuberculosis*. Store at 4 °C and refer to manufacturer's instructions for shelf life.
5. *PTx*: PTx is supplied in the form of 50 μg of lyophilized powder in a sealed vial. Prepare the stock PTx by adding 500 μL of sterile water into the vial and mixing gently until the full contents have dissolved. This results in a concentration of 100 μg/mL. This preparation can be stored at 4 °C in the original vial, for at least 6 months (*see* **Note 2**).
6. *$MOG_{35–55}$ peptide* for the C57BL/6J and NOD/ShiLtJ mouse strains: MEVGWYRSPFSRVVHLYRNGK (*see* **Note 3**) [57, 58].
7. *$PLP_{180–199}$ peptide* for the BALB/cJ mouse strain: WTTCQSI AFPSKTSASIGSL [59].
8. *Stock peptide preparations*: Peptides are most commonly supplied in the form of lyophilized powders. The required amount per mouse is 200 μg. A practical approach for long-term storage is to aliquot amounts sufficient for cohorts of 20 mice. For each aliquot, weigh out 4 mg of peptide using a clean, dry spatula and a fine balance, in a sterile cryogenic tube of 2 mL capacity and store at −20 °C (*see* **Note 4**).

3 Methods

3.1 Preparation of Reagents for EAE Induction

The induction procedure requires two components, namely, (a) an emulsion of peptide in CFA (*see* **Note 5**) supplemented with 4 mg/mL of *M. tuberculosis* (Fig. 2a) and (b) PTx (Fig. 2b). These should be prepared only in amounts required and immediately before use, to avoid long storage times. There is considerable individual variation in clinical parameters, such as timing of clinical onset, disease severity, and survival time; therefore, the use of freshly prepared reagents removes one variable.

1. Spatula, cleaned and dried.
2. Disposable polystyrene weighing trays (55 × 35 mm).
3. Pestle and mortar (9 cm in diameter) cleaned and dried (Fig. 3).
4. Parafilm sealing film.
5. Battery-powered mini-mixer and sterilized homogenizers (Fig. 4a).
6. Sterile, disposable tubes of 5.0 mL capacity (Fig. 4b).
7. Sterile, disposable syringes of 1.0 mL capacity.
8. Sterile, disposable 25-gauge needles.

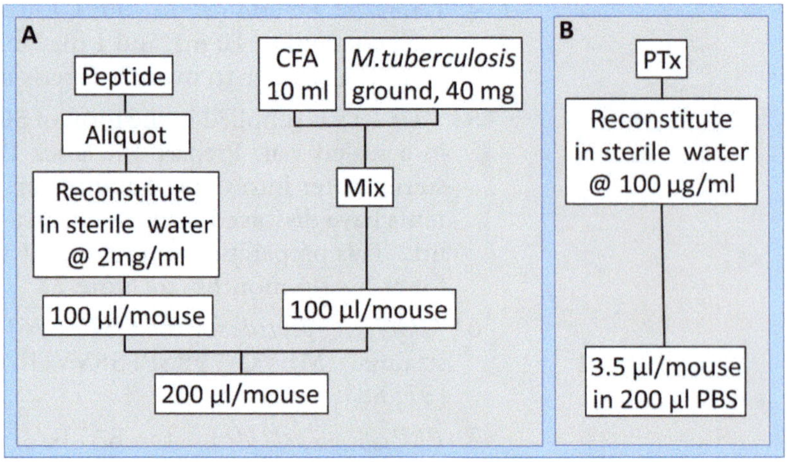

Fig. 2 Preparation of components of the induction procedure. EAE induction requires two components, namely, an emulsion of peptide in *M. tuberculosis*-supplemented CFA (**a**) and PTx (**b**). (**a**) The peptide is reconstituted in sterile water at the concentration of 2 mg/mL, while CFA is supplemented with *M. tuberculosis* at 4 mg/mL. A 1:1 mixture of 100 μL of each component yields the recommended amount of 200 μg of peptide in 200 μL of emulsion/mouse. (**b**) PTx is reconstituted separately in sterile water at a concentration of 100 μg/mL. A 3.5 μL volume yields the 350 ng amount required per mouse and is delivered in 200 μL of PBS. In the case of sham-injected mice, the peptide preparation is substituted with PBS

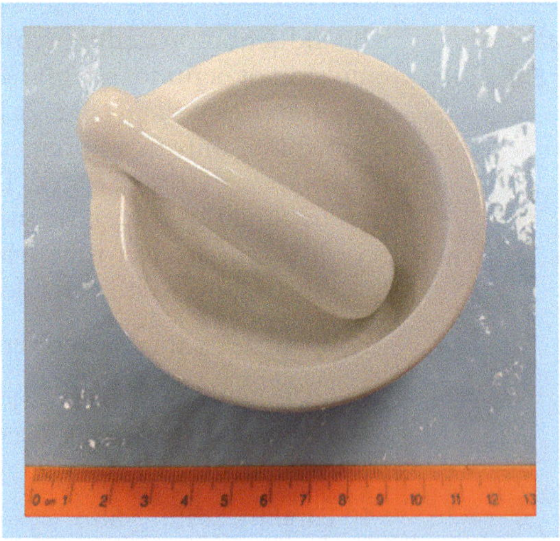

Fig. 3 Utensil for the preparation of finely ground *M. tuberculosis*. The grinding of heat-inactivated *M. tuberculosis* prior to supplementation of CFA is an essential step; otherwise, the bacterium will form clumps when added to CFA resulting in variable dosage in experimental mice. This can be achieved with the use of a small mortar and pestle, 9 cm in diameter, without major loss of material on the wall of the mortar

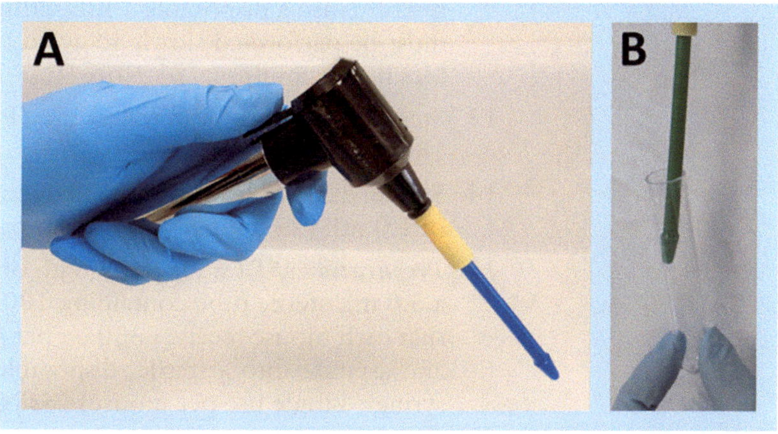

Fig. 4 Preparation of the peptide/*M. tuberculosis*-supplemented CFA emulsion. A convenient way to prepare the peptide in *M. tuberculosis*-supplemented CFA emulsion is to use a small battery-powered mini-mixer (**a**) and a 5 mL tube (**b**). The tube can hold a sufficient amount for five mice without spillage

9. *Finely ground M. tuberculosis.* To prepare *M. tuberculosis* for supplementation of CFA, break the ampule, and using a clean, dry spatula, weigh out 40 mg of heat-inactivated *M. tuberculosis* into a small weighing tray. Reseal the ampule with parafilm sealing film and return it to 4 °C. Transfer the whole amount of *M. tuberculosis* to a small mortar. Using the pestle, thoroughly manually grind into a fine powder for about 15–20 min. Carefully scrape all the fine powder from the walls of the mortar and transfer it back into the weighing tray (*see* **Note 6**).
10. *CFA supplemented with heat-inactivated M. tuberculosis*: Transfer the 40 mg of finely ground *M. tuberculosis* from **step 9**, directly into a single vial of CFA. The resulting *M. tuberculosis* concentration will be 5 mg/mL (*see* **Note** 7). Mix vigorously to obtain a fine suspension of *M. tuberculosis* prior to use.
11. *Preparation of peptide:* On the day of disease induction, add 2 mL of deionized water to a single cryogenic tube containing lyophilized peptide. Mix by gentle repeated inversion until completely dissolved. Pipette 100 μL of peptide solution per mouse. Any remaining solubilized peptide can be frozen and stored at −20 °C. Avoid multiple freeze–thaw cycles.
12. *Peptide/CFA emulsion*: For each mouse, combine 100 μL of *M. tuberculosis*-supplemented CFA from step 10 and 100 μL (200 μg) of peptide from **step 11**, into a 5.0 mL sterile tube. This results in a total of 200 μL of emulsion/mouse. Mix vigorously, using a battery-powered mini-homogenizer, for about 30–45 s (or as long as required to obtain a stable emulsion). Attach a 25-gauge needle to the syringe and aspirate the

mixture into a disposable 1 mL syringe (*see* **Note 8**). This step must be performed slowly to avoid the generation of air bubbles in the emulsion (*see* **Note 9**).

13. Sterile phosphate-buffered saline (PBS): 0.01 M phosphate, 150 mM sodium chloride, pH 7.4.
14. Sterile, disposable, Terumo U-100 insulin syringes with needle of 1.0 mL (27-gauge × 13 mm) capacity.
15. *Preparation of PTx*: Pipette 3.5 μL of stock PTx and transfer to a 5.0 mL sterile tube containing 100–200 μL of sterile PBS so that each mouse receives 350 ng per injection. Mix thoroughly and aspirate into a sterile, disposable, Terumo U-100 insulin syringe (single use per mouse) (*see* **Note 10**).
16. Rodent restrainer of suitable dimensions for a 20–30 g mouse.

3.2 3.2 Final Concentrations of Components/Mouse

Peptide: 200 μg

CFA: 100 μL, containing 500 μg *M. tuberculosis*

PBS: 100 μL

PTx: 350 ng

3.3 Preparation of Reagents for Sham Induction

For sham-injected control mice, repeat **steps 1–16**, except for the substitution of antigenic peptide with saline solution.

3.4 EAE Induction

Prior to performing EAE induction, personnel must be fully trained in the techniques of subcutaneous (s.c.), intraperitoneal (i.p.), and intravenous (i.v.) injections, in monitoring of experimental animals for signs of disease and in humane killing of mice. Grading of disease severity is achieved using a clinical scoring system (Fig. 5). This is based on visual assessment of ambulatory difficulties which result from accumulation of spinal cord lesions and progress in a caudo-rostral direction. The same criteria apply to all three mouse strains. Because of the invasive nature of the protocols and severity of disease, approval must be obtained from the institutional animal ethics committee.

EAE induction is a two-step process taking place over 48 h. On day 0, mice receive a s.c. injection of peptide/*M. tuberculosis*-supplemented CFA emulsion (*see* **Note 11**) and a first PTx injection. The PTx injection is delivered via either the i.p. or i.v. route, depending upon the mouse strain (*see* **Note 12**). At 2 days post induction (dpi), mice receive a second PTx injection containing the same dose as the first (Fig. 6).

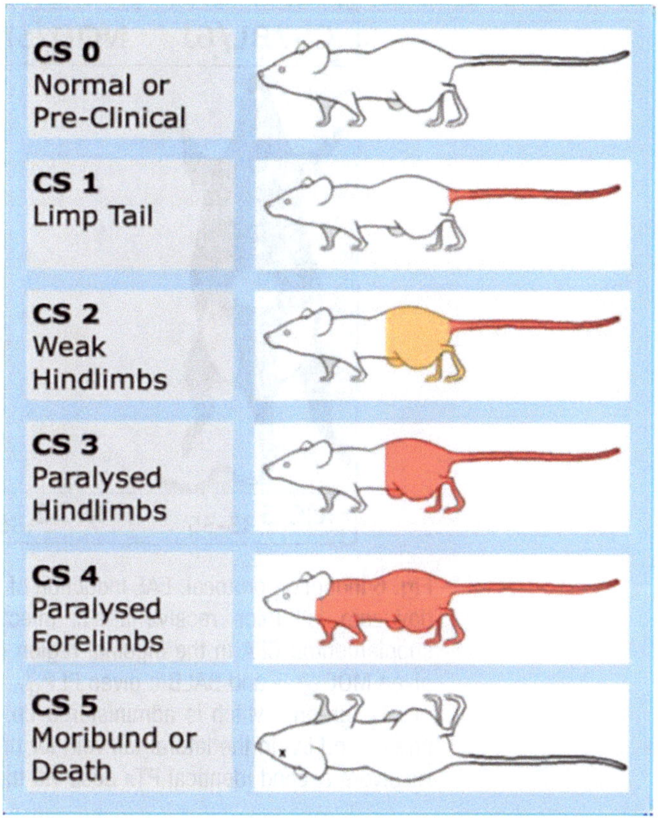

Fig. 5 Clinical scoring of EAE-induced mice. EAE is monitored using the clinical scoring system shown. This is based on visual assessment of the degree of ambulatory difficulties, which is a measure of ascending paralysis due to spinal cord inflammation. Scores range from 0 for no sign of disease up to a score of 5 indicative of a moribund state. A value of 0.5 can be added to the score, where intermediate symptoms are observed. Mice are euthanized at the maximum score of 4.0. Weakness = yellow; paralysis = red

3.5 Generation of a Chronic-Progressive EAE Variant in the C57BL/6J Mouse with MOG_{35-55} Peptide [60] (Fig. 7) (See Notes 13–15)

1. Restrain the mouse in the palm of one hand by holding it by the scruff (thumb and first finger) and tail (third finger) with its abdomen facing up.
2. Inject 100 μL of MOG_{35-55} peptide/*M. tuberculosis*-supplemented CFA emulsion in each inguinal region, by the s.c. route. Pause briefly before gently withdrawing the needle. Wipe off any leakage of emulsion.
3. Immediately follow with an i.p. injection of PTx.
4. Repeat the procedure for all animals using fresh needles each time.
5. Monitor for any bleeding or signs of distress for 30 min post-injection.
6. Repeat the i.p. PTx injection at 2 dpi.

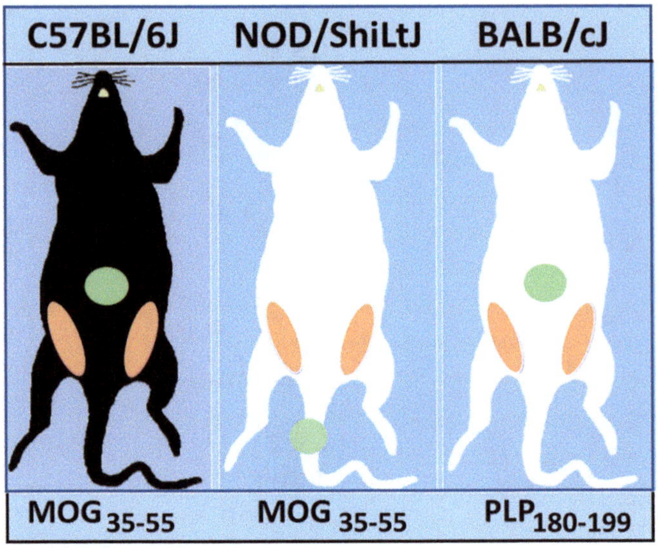

Fig. 6 Induction protocol. EAE induction of C57Bl/6, NOD/ShiLt, and BALB/c. On day zero, all mice receive a s.c. injection of peptide in *M. tuberculosis*-supplemented CFA in the inguinal region (yellow) with C57Bl/6 and NOD/ShiLt given MOG_{35-55} and BALB/c given $PLP_{180-199}$. All mice also receive a first dose of PTx (green), which is administered i.p. for the C57Bl/6 and BALB/c mouse strains and i.v. in the lateral tail vein for the NOD/ShiLt strain. At 2 dpi, all mice receive a second identical PTx dose via the same route as the first

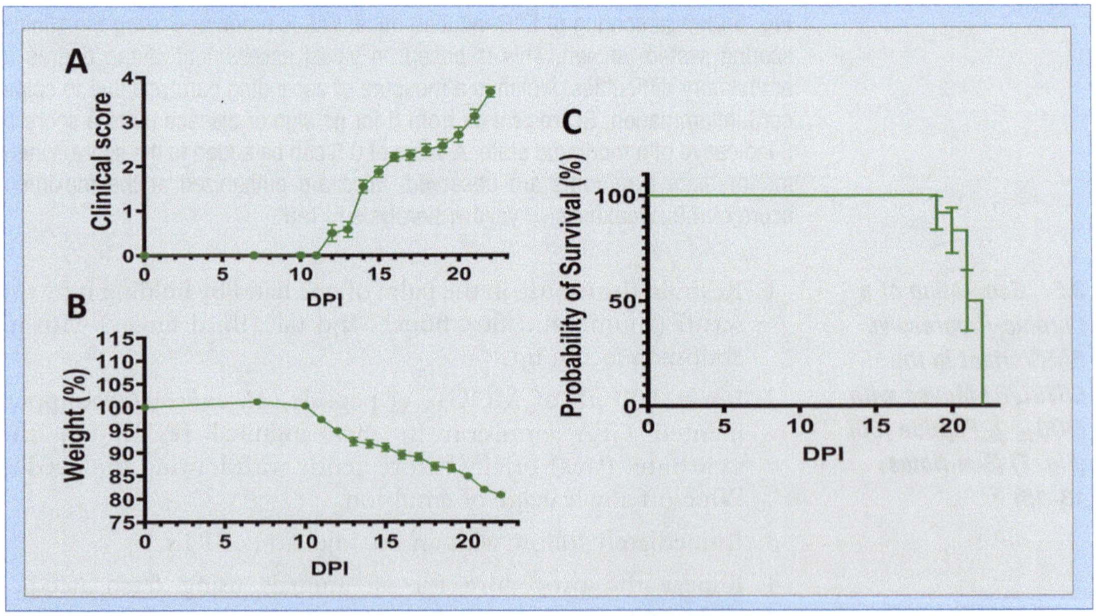

Fig. 7 Generation of a chronic-progressive EAE variant in the C57BL/6J mouse with MOG_{35-55} peptide. In the C57BL/6 mouse strain, the administration of 200 μg of MOG_{35-55} peptide in *M. tuberculosis*-supplemented CFA, together with two doses of 3.5 ng (each) of PTx, delivered via the i.p. route, generates an aggressive profile characterized by clinical onset at about 12–14 dpi and chronic disease without remissions (**a**). Disease severity worsens rapidly and is associated with severe weight loss (**b**). Ethical end point, with >96% of mice requiring humane killing, is reached by 18–22 dpi (**c**). $N = 12$

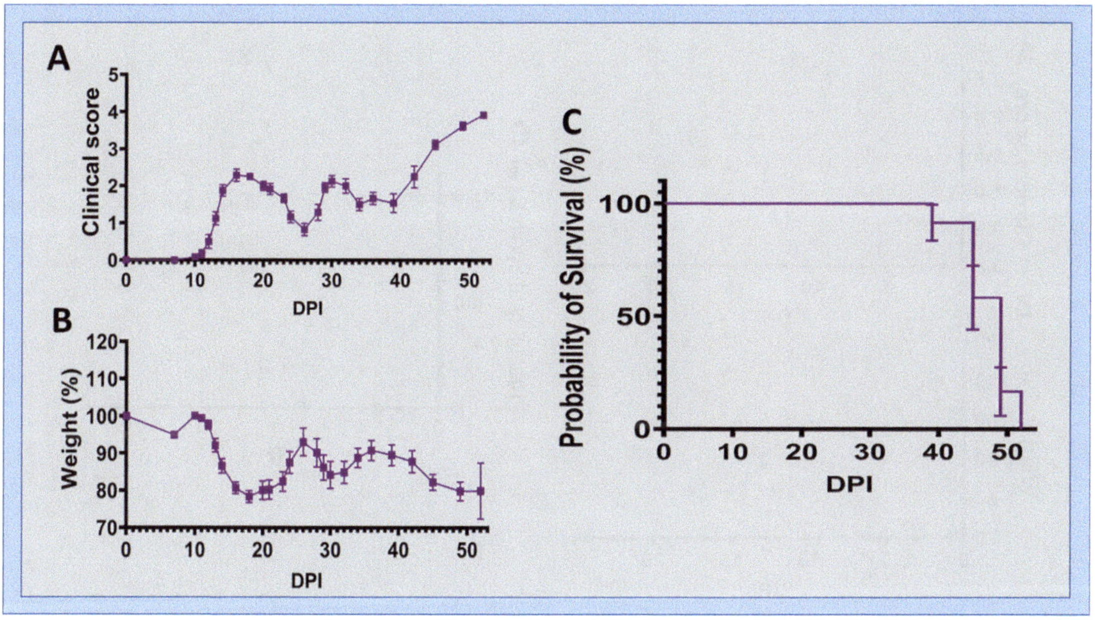

Fig. 8 Generation of a chronic-relapsing EAE variant in the NOD/LtJ mouse with $MOG_{35–55}$ peptide. In the NOD/LtJ mouse strain, the administration of 200 μg of $MOG_{35–55}$ peptide in *M. tuberculosis*-supplemented CFA, together with two doses of 3.5 ng (each) of PTx, delivered via the i.v. route, generates a relapsing–remitting profile characterized by clinical onset at about 11–12 dpi. The first attack is usually synchronized and of short duration (4–6 days) and is followed by almost complete remission by 20–22 dpi (**a**). Subsequent attacks are less predictable and can number between 0 and 4, with incomplete recovery. Attacks are associated with severe weight loss and remissions with weight gain (**b**). Ethical end point, with >96% of mice requiring humane killing, is reached by 40–50 dpi (**c**). $N = 12$

3.6 Generation of a Chronic-Relapsing EAE Variant in the NOD/ShiLtJ Mouse with $MOG_{35–55}$ Peptide [61] (Fig. 8) (See Notes 16 and 17)

1. Restrain the mouse in the palm of one hand by holding it by the scruff (thumb and first finger) and tail (third finger) with its abdomen facing up.
2. Inject 100 μL of $MOG_{35–55}$ peptide/*M. tuberculosis*-supplemented CFA emulsion in each inguinal region, by the s.c. route. Pause briefly before gently withdrawing the needle. Wipe off any leakage of emulsion.
3. Place the mouse in a rodent restrainer.
4. Perform the PTx injection via the i.v. route in the lateral tail vein.
5. Repeat the procedure for all animals using fresh needles each time.
6. Monitor for any bleeding or signs of distress for 30 min post-injection.
7. Repeat the i.v. PTx injection at 2 dpi, on the opposite side on the tail.

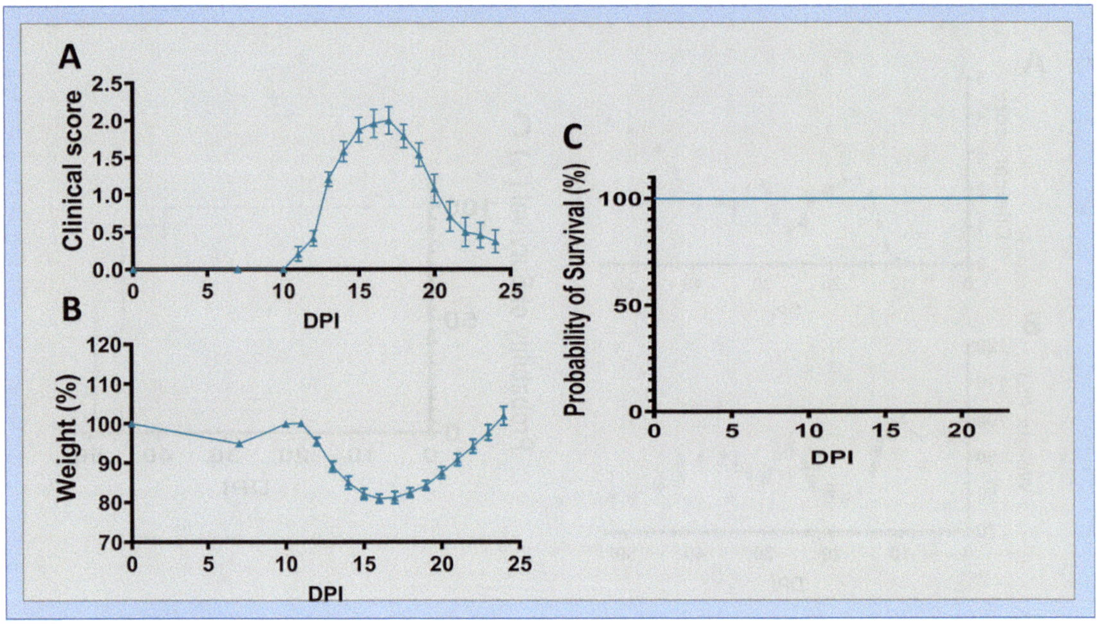

Fig. 9 Generation of a monophasic EAE variant in the BALB/cJ mouse with $PLP_{180–199}$ peptide. In the BALB/cJ mouse strain, the administration of 200 μg of $MOG_{180–199}$ peptide in *M. tuberculosis*-supplemented CFA, together with two doses of 3.5 ng (each) of PTx, delivered via the i.p. route, generates a mild disease profile characterized by clinical onset at about 11–12 dpi and monophasic disease, with disease resolution by about 20–25 dpi (**a**). Attacks are associated with severe weight loss and remissions with weight gain (**b**). Complete recovery is observed in >98% of mice (C). $N = 12$

3.7 Generation of a Monophasic EAE Variant in the BALB/cJ Mouse with $PLP_{180–199}$ Peptide [62] (Fig. 9) (See Notes 18 and 19)

1. Restrain the mouse in the palm of one hand by holding it by the scruff (thumb and first finger) and tail (third finger) with its abdomen facing up.
2. Inject 100 μL of $PLP_{180–199}$ peptide/*M. tuberculosis*-supplemented CFA emulsion in each inguinal region, by the s.c. route. Pause briefly before gently withdrawing the needle. Wipe off any leakage of emulsion.
3. Immediately follow with an i.p. injection of PTx.
4. Repeat the procedure for all animals using fresh needles each time.
5. Monitor for any bleeding or signs of distress for 30 min post-injection.
6. Repeat the i.p. PTx injection at 2 dpi.

3.8 Monitoring of Disease Progression and Care of Animals

1. Mice should be housed at 23 °C, 40% humidity, and 12:12 h light/dark cycle, on standard rodent chow and autoclaved tap water ad libitum (see **Note 20**).

2. Following transport from the breeding facility, allow mice to acclimatize for at least 1 week prior to disease induction (see **Note 21**).
3. Immediately following induction, keep the animals under observation for at least 30 min to ensure that no adverse reaction occurs.
4. Subsequently, check the mice at 2, 7, and 10 dpi, by visual observation and weighing with clinical scores and weights recorded on monitoring sheets. Inspect the injection site for potential skin abrasions. If the mice develop any abrasion, seek veterinary advice.
5. Immediately following the induction both experimental and sham-injected mice exhibit minor weight loss of ≤5%. Verify that restoration of preinduction weight is observed after 1 week (*see* **Note 22**).
6. From day 10 onward, increase monitoring to daily weighing and scoring of disease symptoms.
7. From clinical score 2, food and water must be placed at the bottom of the cage. This should be initiated as soon as this score is reached, because the disease can progress rapidly. Additionally, provide soft material for nest building, non-wetting water gel for hydration (such as HydroGel® which consists of 98% water) and soft food at the bottom of the cage. This must be replaced/replenished when the mice are checked (*see* **Note 23**).
8. From clinical score 3, check mice twice a day.
9. Animals with a score of 4 or that have lost 20% of their baseline weight cannot be left overnight and need to be humanely killed by the end of the day.

4 Notes

1. To ensure the highest disease incidence, mice need to have reached physical maturity. Cohorts consisting of mice at a minimum of 9 weeks of age and weighing a minimum of 21 g at baseline exhibit ≥96% disease incidence. Charts of weight versus age for most common mouse strains are available on the Jackson Laboratory website, under the phenotype information.
2. Bearing in mind that PTx is a protein, storage time should be minimized and the vial kept on ice when removed from the refrigerator.
3. The purity of peptide preparations should be ≥96%. Lower purity results in reduced disease incidence and severity.

4. The most effective way to prevent peptide degradation is to store the lyophilized powder at −20 °C. For reconstituted peptides, recommended concentrations range between 1 and 10 mg/mL and recommended storage temperature is −20 °C.
5. The use of CFA is essential. IFA alone not only fails to induce EAE but appears to be actively tolerogenic. This is so, because IFA apparently induces a Th2 response, as opposed to the Th1 response induced by CFA [44]. There are some reports of successful generation of EAE in the absence of complete adjuvant; however, these resulted in extremely mild disease [38].
6. Finely ground *M. tuberculosis* should be prepared immediately before addition to CFA. It is essential that the *M. tuberculosis* preparation should be finely ground; otherwise, clumping will result upon addition to CFA, which will have an impact on disease incidence and clinical profile. *M. tuberculosis*-supplemented CFA can be stored at 4 °C for at least 12 months.
7. Experience has shown that a minimum of 4 mg/mL CFA is critical for disease development. Lower concentrations severely impact on disease amplitude [30].
8. Because a certain amount of the emulsion will remain in the tip of the syringe and hub of the needle and to minimize the loss of costly reagents, it is recommended to mix quantities of emulsion components sufficient for batches of five mice and load the full amount in a 1.0 mL syringe. However, in practice, due to inevitable loss of some material, this amount will be sufficient for only four mice.
9. The emulsion can be stored at 4 °C for at least 7 days. Where storage of the emulsion is required, it is necessary to ensure that no separation of components has occurred over this period of time. If this appears to be the case, the emulsion can be reconstituted by repeated transfer of the mixture between two syringes connected via a three-way tap.
10. PTx is a critical component for the generation of most EAE variants. Additionally, most commonly two doses, separated by 48 h, are required. Omission of one or both doses will severely impact disease incidence.
11. The emulsion is always delivered by s.c. injection. It is important to use multiple sites to reduce the risk of skin abrasions caused by *M. tuberculosis*, if some of the emulsion spills onto the skin. Injections in the inguinal region can cause a limp which may lead to artificially augmented clinical scores. Some investigators perform injections in alternate sites, for example, two on either side of the midline on the lower back and one on the midline of the back just below the shoulders.

12. Note that in EAE variants that require intravenous delivery of PTx, the maximum recommended injection volume for a dosing solution that is given rapidly is 1 mL/kg body weight (equivalent to 200 μL for a 20 g mouse) for most laboratory animal species, to reduce the risk of hemodilution [63].
13. Considerations when using this variant: The C57BL/6 mouse is the most commonly used one in the generation of EAE. It was originally generated for investigations of immunology and antitumor activity [64]. It is robust and easily bred and, as mentioned, extensively used in the production of genetically modified mutants. It exhibits a predominantly Th1-type autoimmune response. It was the second mammalian species to see its complete genome being sequenced (after humans), the first mouse strain to be flown into space [65] and is the most advanced in functional analysis of its genes. It has low incidence of spontaneous tumors.
14. For the C57BL/6 mouse strain, a range of alternative protocols resulting in various disease profiles, by modulation of MOG_{35-55} and PTx dosage, as well as CFA formulation have been published [66].
15. The disease generated using the above protocol results in a chronic progressive clinical profile (Fig. 7). The earliest symptom, namely, a sudden weight loss of about 10%, is observed around 11–12 dpi, followed by loss of tail tonicity and ambulatory difficulties from 12 to 14 dpi. Symptoms increase in severity over the next week to 10 days and ethical end point is usually reached around 21–22 dpi. Disease incidence using the above protocol is >96%.
16. Considerations when using this variant: The chronic-relapsing clinical profile of this variant offers the major advantage of facilitating investigations of immunopathological mechanisms over the chronic phase and disease resolution. Note that the NOD/ShiLtJ mouse line has a predisposition to type 1 diabetes [62]. Onset of diabetes without remission manifests between 14 and 18 weeks of age, with a higher incidence in females (80–100%, by 30 weeks of age) compared with males (40–60% at the same age). However, this condition is inhibited by CFA; therefore, diabetes is not a confounding issue when NOD/-ShiLtJ mice are used for EAE [67].
17. The disease generated using the above protocol results in a chronic-relapsing clinical profile, with the timing of disease onset similar to that of C57BL/6 mice (Fig. 8). The first attack is synchronized and moderately severe, with a peak clinical score ≤3.0 at 17–18 dpi. This is followed by rapid partial to complete remission to clinical score of 0–1 by 20–22 dpi. Subsequently, the disease course is less predictable, with the

second attack occurring around 25–35 dpi. The subsequent number of attacks and severity of each attack per mouse are highly variable. In our experience, most mice (80%) will experience 2–3 more attacks before reaching ethical end point. However, some will experience no further attacks or in extreme cases additional 4–5 attacks, and it is not uncommon for some mice to reach ethical end point between 60 and 150 dpi. The disease incidence at the first attack is >96%.

18. Considerations when using this variant: The BALB/c mouse is frequently used in studies of inflammation and autoimmunity. It exhibits a predominantly Th2-type autoimmune response and has a stronger humoral response compared with C57BL/6 mice. The rate of spontaneous tumors increases with age, especially lung and kidney neoplasia. The disease generated using the above protocol results in monophasic clinical profile, with the timing of disease onset similar to that of variants described above (Fig. 9).
19. In this variant, only about 60–66% of mice exhibit overt disease, with maximum clinical scores ranging from 1 to 4. However, our laboratory has shown that in actual fact almost 100% of mice develop lesions. The reason for the absence of overt disease is that a proportion of animals develop lesions only in the upper CNS and do not develop lesions in the spinal cord, hence the absence of clinical scores [62].
20. EAE will manifest in a conventional or SPF environment. However, disease severity is reduced with increasing sterility of environment; therefore, a germ-free facility is not suitable [68].
21. Stress can negatively affect disease development and should be avoided. This is related to increased levels of corticosterone which have been shown to inhibit EAE [30, 67].
22. Sham-injected animals do not exhibit further symptoms and continue to gain weight as expected for their strain.
23. Compared with other mouse strains, NOD/ShiLtJ mice require more rigorous surveillance as they seem to consume less food and water than other mice once they reach clinical score 3.

Acknowledgments

The authors thank La Trobe University for research scholarships to DL Maxwell and VJT Lim and La Trobe Animal Research and Teaching Facility (LARTF) for technical support. The work was funded by private donations and grants from La Trobe Research Focus Areas and Multiple Sclerosis Research Australia (MSRA).

References

1. Baker D, Gerritsen W, Rundle J, Amor S (2011) Critical appraisal of animal models of multiple sclerosis. Mult Scler J 17:647–757
2. Rivers YM, Schwentker FF (1935) Encephalomyelitis accompanied by myelin destruction experimentally produced in monkeys. J Exp Med 61:689–702
3. Baxter AG (2007) The origin and application of experimental autoimmune encephalomyelitis. Nature 7:904–914
4. Mix E, Meyer-Rienecker H, Hartung H-P, Zettl UK (2010) Animal models of multiple sclerosis- potentials and limitations. Prog Neurobiol 92:386–404
5. Kipp M, Nyamoya S, Hochstrasser T, Amor S (2017) Multiple sclerosis animal models: a clinical and histopathological perspective. Brain Pathol 27:123–137
6. Bjelobaba I, Begovic-Kupresanin V, Pekovic S, Lavrnja I (2018) Animal models of multiple sclerosis: focus on experimental autoimmune encephalomyelitis. J Neuro Res 96:1021–1042
7. Genain CP, Hauser SL (1997) Experimental allergic encephalomyelitis in the New World monkey Callithrix jacchus. Immunol Rev 183: 159–172
8. 't Hart BA, van Kooyk Y, Geurts JJ, Gran B (2015) The primate autoimmune encephalomyelitis model; a bridge between mouse and man. Ann Clin Transl Neurol 2:581–593
9. Kap YS, Jagessar SA, Dunham J, 't Hart BA (2016) The common marmoset as an indispensable animal model for immunotherapy development in multiple sclerosis. Drug Discov Today 21:1200–1205
10. Pollak Y, Orion E, Goshen I et al (2002) Experimental autoimmune encephalomyelitis-associated behavioral syndrome as a model of 'depression due to multiple sclerosis'. Brain Behav Immun 16:533–543
11. Haji N, Mandelosi G, Gentile A et al (2013) TNF-α-mediated anxiety in a mouse model of multiple sclerosis. Exp Neurol 237:296–303
12. Olechowski C, Tenorio G, Sauve Y, Kerr BJ (2013) Changes in nociceptive sensitivity and object recognition in experimental autoimmune encephalomyelitis (EAE). Exp Neurol 241:113–121
13. Bonfiglio T, Olivero G, Merega E et al (2017) Prophylactic versus therapeutic fingolimod: restoration of presynaptic defects in mice suffering from experimental autoimmune encephalomyelitis. PLoS One 12:e0170825
14. Kocovski P, Jiang X, D'Souza CS et al (2019) Platelet depletion is effective in ameliorating anxiety-like behavior and reducing the pro-inflammatory environment in the hippocampus in murine experimental autoimmune encephalomyelitis. J Clin Med 8:162–178
15. Kocovski P, Tabassum-Sheikh N, Marinis S et al (2021) Immunomodulation eliminates inflammation in the hippocampus in experimental autoimmune encephalomyelitis, but does not ameliorate anxiety-like behavior. Front Immunol 12:639650
16. Boeschoten RE, Braamse AMJ, Beekman ATF et al (2017) Prevalence of depression and anxiety in Multiple Sclerosis: a systematic review and meta-analysis. J Neurol Sci 372:331–341
17. Rossi S, Studer V, Motta C et al (2017) Neuroinflammation drives anxiety and depression in relapsing-remitting multiple sclerosis. Neurology 89:1338–1347
18. Lassmann H (2019) Pathogenic mechanisms associated with different clinical courses of multiple sclerosis. Front Immunol 9:3116
19. Ayers MM, Hazelwood LJ, Catmull DV et al (2004) Early glial responses in murine models of multiple sclerosis. Neurochem Int 45:409–419
20. Wang D, Ayers MM, Hazelwood LJ et al (2005) Astrocyte-associated axonal damage in pre-onset stages of experimental autoimmune encephalomyelitis. Glia 51:235–240
21. Huizinga R, Gerritsen W, Heijmans N, Amor S (2008) Axonal loss and grey matter pathology as a direct result of autoimmunity. Neurobiol Dis 32:461–470
22. Gold R, Linington C, Lassmann H (2006) Understanding pathogenesis and therapy of multiple sclerosis via animal models: 70 years of merits and culprits in experimental autoimmune encephalomyelitis research. Brain 129: 1953–1971
23. Baker D, Amor S (2014) Experimental autoimmune encephalomyelitis is a good model of multiple sclerosis if used wisely. Mult Scler Rel Disord 3:555–564
24. Stromnes IM, Goverman JM (2006) Passive induction of experimental autoimmune encephalomyelitis. Nat Protoc 1:1952–1960
25. Sun D, Whitaker JN, Huang Z et al (2001) Myelin antigen-specific $CD8^+$ T cells are encephalitogenic and produce severe disease in C57BL/6 mice. J Immunol 166:7579–7987
26. Terry RL, Ifergan I, Miller SD (2016) Experimental autoimmune encephalomyelitis in mice. Method Mol Biol 1304:145–160

27. Glatigny S, Betelli E (2018) Experimental autoimmune encephalomyelitis (EAE) as animal models of multiple sclerosis (MS). Cold Spring Harb Perspect Med 8:a028977
28. Krishnamoorthy G, Lassmann H, Wekerle H, Holz A (2006) Spontaneous opticospinal encephalomyelitis in a double-transgenic mouse model of autoimmune T cell/B cell cooperation. J Clin Invest 116:2385–2392
29. Kuerten S, Angelov DN (2008) Comparing the CNS morphology and immunobiology of different EAE models in C57BL/6 mice – a step towards understanding the complexity of multiple sclerosis. Ann Anat 190:1–15
30. Miller SD, Karpus WJ, Davidson TS (2007) Experimental autoimmune encephalomyelitis in the mouse. Curr Protoc Immunol Chapter 15:15.1.1–15.1.18
31. Storch MK, Stefferl A, Brehm U et al (1998) Autoimmunity to myelin oligodendrocyte glycoprotein in rats mimics the spectrum of multiple sclerosis pathology. Brain Pathol 8:681–694
32. Emerson MR, Gallagher RJ, Marquis JG, Le Vine SM (2009) Enhancing the ability of experimental autoimmune encephalomyelitis to serve as a more rigorous model of multiple sclerosis through refinement of experimental design. Comp Med 59:112–128
33. Weissert R, Wallstrom E, Storch M et al (1998) MHC haplotype-dependent regulation of MOG-induced EAE in rats. J Clin Invest 102: 1265–1273
34. Wang L, Winnewisser J, Federle C et al (2017) Epitope-specific tolerance modes differentially specify susceptibility to proteolipid protein-induced experimental autoimmune encephalomyelitis. Front Immunol 8:1511
35. Bernard C, Johns TG, Slavin A et al (1997) Myelin oligodendrocyte glycoprotein: a novel candidate autoantigen in multiple sclerosis. J Mol Med 75:77–88
36. Sobel RA, van der Veen RC, Lees MB (1986) The immunopathology of chronic experimental allergic encephalomyelitis induced in rabbits with bovine proteolipid protein. J Immunol 136:157–163
37. Martinsen V, Kursula P (2022) Multiple sclerosis and myelin basic protein: insights into protein disorder and disease. Amino Acids 54: 99–109
38. Stocic-Grujicic S, Ramic Z, Bumbasirevic V et al (2004) Induction of experimental autoimmune encephalomyelitis in Dark Agouti rats without adjuvants. Clin Exp Immunol 136: 49–55
39. Muller DM, Pender MP, Greer JM (2000) A neuropathological analysis of experimental autoimmune encephalomyelitis with predominant brainstem and cerebellar involvement and differences between active and passive induction. Acta Neuropathol 100:174–182
40. Kerlero de Rosbo N, Hoffmann M, Mendel I et al (1997) Predominance of the autoimmune response to myelin oligodendrocyte glycoprotein (MOG) in multiple sclerosis: reactivity to the extracellular domain of MOG is directed against three main regions. Eur J Immunol 27: 3059–3069
41. Croxford AL, Kurschus F, Waisman A (2011) Mouse models for multiple sclerosis: historical facts and future implications. Biochim Biophys Acta 1812:177–183
42. Rangachari M, Kuchroo VK (2013) Using EAE to better understand principles of immune function and autoimmune pathology. J Autoimmun 45:31–39
43. Lee YK, Menezes JS, Umesaki Y, Mazmanian SK (2011) Proinflammatory T-cell responses to gut microbiota promote experimental autoimmune encephalomyelitis. Proc Natl Acad Sci U S A 108:4615–4622
44. Constantinescu CS, Hilliard BA (2005) Adjuvants in EAE. In: Lavi E, Constantinescu C (eds) Experimental models of multiple sclerosis. Springer, New York, pp 73–84
45. Awate S, Babiuk LA, Mutwiri G (2013) Mechanisms of action of adjuvants. Front Immunol 4:114
46. Namer IJ, Steibel J, Poulet P et al (1994) The role of *Mycobacterium tuberculosis* in experimental allergic encephalomyelitis. Eur Neurol 34:224–227
47. Moss J, Stanley SJ, Burns DL et al (1983) Activation by thiol of the latent NAD glycohydrolase and ADP-ribosyltransferase activities of *Bordetella pertussis* toxin (islet-activating protein). J Biol Chem 258:11879–11882
48. Sriram S, Steiner I (2005) Experimental autoimmune encephalomyelitis: a misleading model of multiple sclerosis. Ann Neurol 58: 939–945
49. Ransohoff RM (2012) Animal models of multiple sclerosis: the good, the bad and the bottom line. Nat Neurosci 15:1074–1077
50. Behan PO, Chaudhauri A (2014) EAE is not a useful model for demyelinating disease. Mult Scler Relat Dis 3:565–574
51. Lassmann H, Bradl M (2017) Multiple sclerosis: experimental models and reality. Acta Neuropathol 133:223–244

52. D'haeseleer, Cambron M, Vanopdenbosh L, De Keyser J (2011) Vascular aspects of multiple sclerosis. Lancet Neurol 10:657–666
53. Linden JR, Ma Y, Zhao B et al (2015) *Clostridium perfinges* epsilon toxin causes selective death of mature oligodendrocytes and central nervous system demyelination. mBio 6: e02513–e02514
54. 't Hart BA, Laman JD, Kap YS (2018) Merits and complexities of modeling multiple sclerosis in non-human primates: implications for drug discovery. Exp Opin Drug Discov 13:387–397
55. Lassmann H (2010) Axonal and neuronal pathology in multiple sclerosis: what we have learnt from animal models. Exp Neurol 225:2–8
56. Robinson AP, Harp CT, Noronha A, Miller SD (2014) The experimental autoimmune encephalomyelitis (EAE) model of MS: utility for understanding disease pathophysiology and treatment. Handb Clin Neurol 122:173–189
57. Mendel I, Kerlero de Rosbo N, Ben-Nun A (1995) A myelin oligodendrocyte glycoprotein peptide induces typical chronic experimental autoimmune encephalomyelitis in H-2b mice: fine specificity and T cell receptor V beta expression of encephalitogenic T cells. Eur J Immunol 25:1951–1959
58. Slavin A, Ewing C, Liu J et al (1998) Induction of a multiple sclerosis-like disease in mice with an immunodominant epitope of myelin oligodendrocyte glycoprotein. Autoimmunity 28: 109–120
59. Lyons JA, Ramsbottom MJ, Trotter JL, Cross AH (2002) Identification of the encephalitogenic epitopes of CNS proteolipid protein in BALB/c mice. J Autoimmun 19:195–121
60. Pham H, Doerrbecker J, Ramp AA et al (2011) Experimental autoimmune encephalomyelitis (EAE) in C56BL/6 mice is not associated with astrogliosis. J Neuroimmunol 232:51–62
61. Dang P, Bui K, D'Souza CS, Orian JM (2015) Modelling MS: MOG-induced EAE in the NOD/Lt mouse strain. In AC La Flamme, JM Orian (eds.) Emerging and evolving topics in MS pathogenesis and treatments. Curr Top Behav Neurosci 26:143–177, Springer De
62. Orian JM, Keating P, Downs L et al (2015) Deletion of IL-4Rα in the BALB/c mouse is associated with altered lesion topography and susceptibility to pertussis toxin-induced EAE. Autoimmunity 48:208–221
63. Turner PV, Brabb T, Pekow C, Vasbinder MA (2011) Administration of substances to laboratory animals: routes of administration and factors to consider. J Am Assoc Lab Anim 50: 600–613
64. Song HK, Hwang DY (2017) Use of C57BL/6N mice on the variety of immunological research. Lab Anim Res 33:119–123
65. Andreev-Andrievskiy A, Popova A, Boyle R et al (2014) Mice in Bion-M 1 space mission: training and selection. PLoS One 9:e104830
66. Berard JL, Wolak K, Fournier S, David S (2010) Characterization of relapsing-remitting and chronic forms of experimental autoimmune encephalomyelitis in C57BL/6 mice. Glia 58:434–445
67. Kocovski P, Dang PT, D'Souza CS et al (2018) Differential anxiety-like responses in NOD/ShiLtJ and C57BL/6J mice following experimental autoimmune encephalomyelitis induction and oral gavage. Lab Anim 52:470–478
68. Wolfensohn S, Hawkins P, Lilley P et al (2013) Reducing suffering in experimental autoimmune encephalomyelitis (EAE). J Pharmacol Toxicol 67:169–176
69. Tompkins SM, Padilla J, Dal Canto MC et al (2002) De novo central nervous system processing of myelin antigen is required for the initiation of experimental autoimmune encephalomyelitis. J Immunol 168:4173–4183
70. Tan LJ, Kennedy MK, Miller SD (1992) Regulation of the effector stages of experimental autoimmune encephalomyelitis via neuroantigen-specific tolerance induction. II. Fine specificity of effector T cell inhibition. J Immunol 148:2748–2755
71. Sakai K, Zamvil SS, Mitchell DJ (1988) Characterization of a major encephalitogenic T cell epitope in SJL/J mice with synthetic oligopeptides of myelin basic protein. J Neuroimmunol 19:21–32
72. Amor S, Groome N, Linington C et al (1994) Identification of epitopes of myelin oligodendrocyte glycoprotein for the induction of experimental allergic encephalomyelitis in SJL and Biozzi AB/H mice. J Immunol 153:4349–4356
73. Tsunoda I, Kuang LQ, Theil DJ, Fujinami RS (2000) Antibody association with a novel model for primary progressive multiple sclerosis: induction of relapsing-remitting and progressive forms of EAE in H2s mouse strains. Brain Pathol 10:402–418
74. Greer JM, Sobel RA, Sette A (1996) Immunogenic and encephalitogenic epitope clusters of myelin proteolipid protein. J Immunol 156: 371–379
75. Tuohy VK, Thomas DM (1995) Sequence 104–117 of myelin proteolipid protein is a cryptic encephalitogenic T cell determinant for SJL/J mice. J Neuroimmunol 56:161–170

76. Greer JM, Denis B, Sobel RA, Trifilieff E (2001) Thiopalmitoylation of myelin proteolipid protein epitopes enhances immunogenicity and encephalitogenicity. J Immunol 166: 6907–6913
77. Tuohy VK, Lu Z, Sobel RA et al (1989) Identification of an encephalitogenic determinant of myelin proteolipid protein for SJL mice. J Immunol 142:1523–1527
78. Greer JM, Kuchroo VK, Sobel RA, Lees MB (1992) Identification and characterization of a second encephalitogenic determinant of myelin proteolipid protein (residues 178–191) for SJL mice. J Immunol 149:783–788
79. Fritz RB, Chou CH, McFarlin DE (1983) Induction of experimental allergic encephalomyelitis in PL/J and (SJL/J × PL/J)F1 mice by myelin basic protein and its peptides: localization of a second encephalitogenic determinant. J Immunol 130:191–194
80. Greer JM, Klinguer C, Trifilieff E et al (1997) Encephalitogenicity of murine, but not bovine, DM20 in SJL mice is due to a single amino acid difference in the immunodominant encephalitogenic epitope. Neurochem Res 22:541–547
81. Morris-Downes MM, McCormack K, Baker D et al (2002) Encephalitogenic and immunogenic potential of myelin-associated glycoprotein (MAG), oligodendrocyte-specific glycoprotein (OSP) and 20,30-cyclic nucleotide 30-phosphodiesterase (CNPase) in ABH and SJL mice. J Neuroimmunol 122:20–33
82. Amor S, O'Neill JK, Morris MM et al (1996) Encephalitogenic epitopes of myelin basic protein, proteolipid protein, myelin oligodendrocyte glycoprotein for experimental allergic encephalomyelitis induction in Biozzi ABH (H-2Ag7) mice share an amino acid motif. J Immunol 156:3000–3008
83. Smith P, Heijmans N, Ouwerling B et al (2005) Native myelin oligodendrocyte glycoprotein promotes severe chronic neurological disease and demyelination in Biozzi mice. Eur J Immunol 35:1311–1319

Chapter 16

Immunohistochemical Analysis of the *Drosophila* Larval Neuromuscular Junction

Yichen Sun, Yu Zhao, Travis K. Johnson, and Wei Xie

Abstract

Synapses are specialized junctions between cells that mediate neurotransmission to modify brain activity and body function. Studies on synapse structure and function play an important role in understanding how neurons communicate and the consequences of their dysfunction in neurological disorders. The *Drosophila* larval neuromuscular junction is an excellent model for dissecting the cellular and molecular mechanisms of the synapse, with its large size, accessibility, and well-characterized genetics. This protocol describes the steps required for morphological and immunohistochemical analysis of the *Drosophila* larval neuromuscular junction including its dissection and multiplex labeling of synaptic proteins. This technique can be used to assess the impact of genetic manipulations on synaptic development, integrity, and plasticity, thus providing a valuable tool for probing complex neurological processes in a whole animal system.

Key words *Drosophila* larvae, Neuromuscular junction, Immunostaining, Dissection, Synapse morphology

1 Introduction

Neuromuscular junctions (NMJs), the specialized synapses that form between motor neurons and skeletal muscles, are essential for adjusting behavior in response to environmental stimuli and modulating body function. The *Drosophila* larval glutamatergic neuromuscular junction (NMJ) is a well-established model [1] for investigating cellular and molecular processes that underpin synapse formation, function, and dynamics. It shares remarkable similarities with the excitatory synapses in vertebrate central nervous system (CNS) [2] and comprises both motor innervation and musculature [3] which coordinate the peristaltic movement of the larva. The motor neurons responsible originate in the ventral nerve cord from one of the three thoracic (T1–T3) or eight abdominal

Yichen Sun and Yu Zhao contributed contributed equally to this work.

Sebastian Dworkin (ed.), *Neurobiology: Methods and Protocols*, Methods in Molecular Biology, vol. 2746, https://doi.org/10.1007/978-1-0716-3585-8_16,

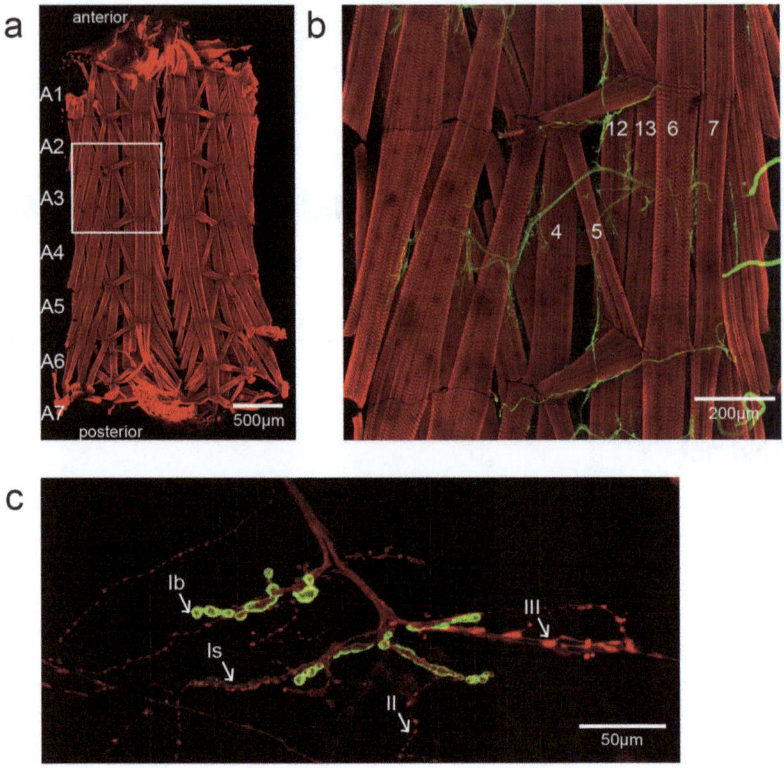

Fig. 1 *Drosophila* body wall muscle and bouton types. (**a**) Body wall muscle of a third-instar larvae stained with phalloidin. Seven abdominal segments are shown here (A1–A7). The white box is enlarged in Fig. 1b. (**b**) The A3 hemi-segment of the *Drosophila* third-instar larvae. This musculature arrangement is invariant in A2–A7. Muscle 4,5,6,7,12, and 13 are labeled. Muscles are stained with phalloidin (red) and NMJs are stained by anti-HRP (green). (**c**) Type Ib, Is, II, and III boutons on muscle 12 are indicated by arrows. Type I boutons exist on every muscle cell and can be further divided into type Ib and Is. With more SSR surrounding, type Ib boutons (3–6 μm diameter) are larger in size than type Is bouton (2–4 μm diameter) [6], which are both glutamatergic. Type II and III boutons completely lack SSR and thus lack the DLG staining signal. Type II boutons have the smallest size (1–2 μm diameter) and use glutamate and/or peptides as neurotransmitters. Type III boutons only exist on muscle 12 in segments A2–A5. They are intermediate in size (2–3 μm diameter) and only use peptides as neurotransmitters. The boutons are labeled with anti-DLG (green) and anti-HRP (red)

(A1–A8) segments of the *Drosophila* larval body, respectively, from anterior to posterior (Fig. 1a). Among these segments, A2–A7 segments share a precise stereotypical repeated muscle pattern, which are ventral midline symmetric. Within these segments (hemi-segments), 30 body wall muscle cells are uniquely defined according to their positions and orientations (Fig. 1b). These are innervated by 32 motor neurons from the corresponding neuromere, each forming NMJs with highly stereotypic morphology.

The *Drosophila* larval NMJs are organized as branched arbors composed of a string of oval-shaped terminals called boutons. There are three types of boutons in *Drosophila* larvae (type I, II, III), with differences in their size, postsynaptic structure subsynaptic reticulum (SSR) characteristics, and neurotransmitter identity [4, 5] (Fig. 1c). Type I boutons are further separated into type Ib and type Is. Type Ib boutons are larger in size than type Is bouton (3–6 μm vs 2–4 μm diameter) [6] and have greater SSR surrounding the axonal terminal of the motor neuron. Both type I boutons utilize glutamate as the neurotransmitter, the dominant excitatory source of excitation in the CNS of vertebrates [7]. Type II and III boutons lack SSR entirely and are consequently much smaller than type I boutons [8]. For simplicity, the present protocol examines only type I boutons, as their relatively large size makes microscopic analysis readily accessible.

Formation of the NMJ occurs during embryonic stages 13–15, and by the end of embryonic development, NMJs are functional [9]. In order to provide sufficient neurotransmission sites for muscle fiber growth, expansion of the *Drosophila* larval NMJ occurs by both bouton enlargement and the addition of new boutons. Since the number of boutons per muscle is relatively constant, it intuitively reflects the growth of NMJ [10]. Efficient neurotransmission relies not only of bouton number but also on the stability and integrity of mature synapse [11]. Synapse development and function are subject to activity-dependent regulatory mechanisms [9] and can be altered by the environment, neural activity, and gene function [12, 13].

The larval NMJ synaptic morphology is routinely examined ex vivo by dissection, fixation, and multiplex immunostaining followed by fluorescence microscopy. This can reveal (sub)cellular structures as well as the organization of the intracellular matrix at individual synapses in intact tissue. Using the technique we describe here in concert with a number of validated primary antibodies and dyes (Table 1), the entire NMJ structure, the number of boutons, as well as the synaptic protein organization can be assessed. These serve as well-established proxies for NMJ growth, complexity, and integrity [10, 11]. When used in conjunction with electrophysiology and behavioral experiments, this technique can provide novel insights into cellular and molecular mechanisms operating at the synapse.

2 Materials

1. Phosphate-buffered saline (PBS): 200 mmol/L Na_2HPO_4, 35 mmol/L KH_2PO_4, 2.74 mol/L NaCl, 53 mmol/L KCl. pH 7.2–7.6.

Table 1
List of primary antibodies and dye, their source, and working dilutions

Antigen/dye	Target (pre- or postsynaptic)	Source	Catalog no.	Dilution	Host species
BRP (bruchpilot)	Active zone (pre)	DSHB[a]	AB_2314866	1:50	Mouse
DLG (discs large)	SSR (post)	DSHB	AB_528203	1:100	Mouse
Fas II (Fasciclin II)	Neuron axon (pre)	DSHB	AB_528235	1:25	Mouse
Futsch	Microtubule in neurons (pre)	DSHB	AB_528403	1:50	Mouse
GluRIIA	Glutamate receptor subunit (post)	DSHB	AB_528269	1:50	Mouse
HRP	Cell membrane of neurons (pre)	JIR[b]	AB_2314648	1:500	Rabbit
Spectrin	Actin-related protein (both)	DSHB	AB_528473	1:25	Mouse
Syn (synapsin)	Synaptic vesicles (pre)	DSHB	AB_528479	1:50	Mouse
Syt (synaptotagmin)	Synaptic vesicles (pre)	DSHB	AB_528483	1:50	Mouse
Phalloidin	F-actin (both)	Invitrogen	R415	1:6–1:500	

[a]DSHB: Developmental Studies Hybridoma Bank
[b]JIR: Jackson ImmunoResearch

2. 4% paraformaldehyde (PFA) in PBS. Store at 4 °C.
3. Methanol. Store at −20 °C.
4. PBST: 0.3% Triton X-100 in PBS.
5. Blocking solution: 1% bovine serum albumin (BSA) in PBST. Store at 4 °C.
6. Primary antibodies (Table 1).
7. Secondary antibodies conjugated with fluorescent dye, such as Alexa Fluor 488.
8. Mounting medium.
9. Two pairs of high-precision straight forceps. Length, 110 mm; tip (width × thickness), 0.05 × 0.01 mm. Suitable for biological use.
10. Minutien pins (6 per larva). Tip diameter, 0.0125 mm; rod diameter, 0.1 mm.
11. Pair of Venus Spring Scissors (6.5 mm cutting edge, curved).
12. A 6 cm-diameter petri dish.
13. Dissection plate: a 6 cm-diameter plate containing a bed of silicone.
14. Glass microscope slides.

15. Cover glass (length × width × thickness): 24 × 32 × 0.13 mm.
16. Clear nail polish.
17. Dissecting microscope (e.g., Zeiss Stemi 305 with 4× eyepiece).
18. Spiral mixer (e.g., spiral mixer KB-3-D from Kylin-Bell Lab Instruments).

3 Methods

3.1 Preparation for Dissection

1. Culture desired strains to allow for the selection of wandering third-instar larvae for dissection.
2. Precool 4% PFA and PBS on ice.

3.2 Tissue Dissection

1. Using forceps, carefully collect a wandering third-instar larvae and place it in the petri dish filled with cold PBS. The cold PBS is used to remove external contaminants such as culture medium and decrease the larval body temperature to cause immobilization.
2. Carefully transfer the larvae with forceps into a drop of PBS placed on the dissection plate. The PBS prevents desiccation of the larvae.
3. To begin the dissection, orient the larvae with its dorsal side up and spiracles down. Pick up a Minutien pin with forceps and carefully pin the larvae through its cuticle between the posterior spiracles and into the dissection plate surface so that it stays in place. Insert another Minutien pin at the anterior near the mouth hooks, using the needle to stretch the animal out lengthwise.
4. With the Venus scissors, cut the larval cuticle across its dorsal width immediately anterior to the posterior pin in such a way that the ventral cuticle (against the plate) remains uncut. Repeat at the anterior end.
5. Using the posterior incision as a starting point, make a third cut along the entire dorsal midline toward the anterior (until this cut connects with the anterior incision). As a guide, try to make the incision between the two parallel dorsal tracheal tubes (*see* **Note 1**).
6. Remove the internal organs. Add several drops of PBS to the larvae to enable the internal organs to float up and out of the body. Carefully remove the viscera, trachea, and the remaining organs with forceps.
7. Use four more Minutien pins to secure the left and right cuticle flaps (top and bottom) in an open position, exposing the

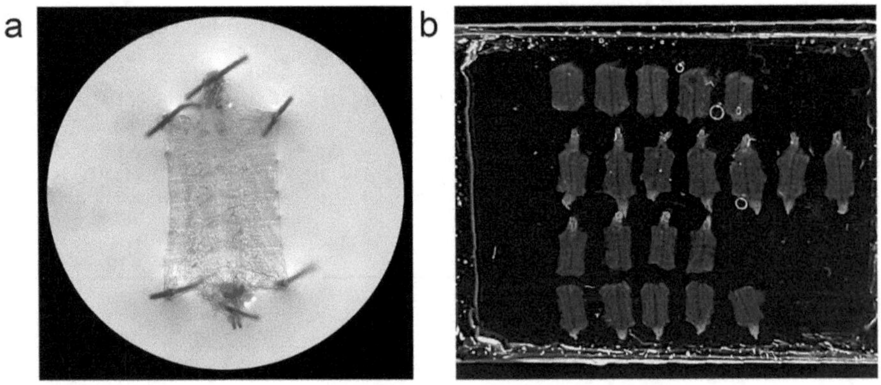

Fig. 2 The *Drosophila* larvae dissection and sample arrangement. (**a**) The dissected *Drosophila* third-instar larvae. Pinned by six Minutien pins on the dissection plate, the specimen is a flattened hexagon shape. (**b**) Samples are arranged according to their genotypes or condition in rows in the same orientation. From top to bottom, the four types of samples are in order, headless and tailless, complete, tailless, and headless

internal musculature and creating a flattened hexagon shape (Fig. 2a). Ensure that the body wall is stretched both in anterior–posterior and left–right directions.

3.3A Standard Fixation

1. After removing the PBS, fix the specimen at room temperature for 40 min in a tissue-covering droplet of precooled 4% PFA (*see* **Note 2**).

3.3B Alternative Fixation (For Membrane Clustered Proteins)

1. For target antigens that cluster in synaptic membranes, such as bruchpilot (BRP) and glutamate receptor (GluR), fix the specimen at room temperature for 5 min in a droplet of precooled methanol.

3.4 Immunostaining

1. Remove the Minutien pins and transfer the samples into a 1.5 mL tube containing 1 mL PBS (*see* **Note 3**).
2. After all the samples have been collected, remove the PBS, and wash the samples on spiral mixer at room temperature with 1 mL PBST, three times for 10 min each, to remove the residual fixative and allow further tissue permeability (*see* **Note 4**).
3. Incubate the samples in 1 mL blocking solution for 1 h on spiral mixer at room temperature to minimize nonspecific antibody binding.
4. Dilute the desired primary antibodies in 1 mL blocking solution and incubate the samples gently shaking overnight at 4 °C (*see* **Note 5**).
5. The following day, wash the samples with 1 mL PBST three times for 10 min each to remove unbound primary antibodies (*see* **Note 6**).

6. Dilute the secondary antibodies conjugated with fluorescent dye in 1 mL blocking solution, and add it to the samples. Incubate for 2 h on spiral mixer at room temperature in the dark.
7. Wash the samples on spiral mixer at room temperature with 1 mL PBST, three times for 10 min each, to remove unbound secondary antibodies.
8. For samples where stain by dyes is desired, dilute the dye in 1 mL blocking solution and add it to the samples. Incubate for the duration recommended by the manufacturer on spiral mixer at room temperature in the dark.
9. Wash the samples on spiral mixer at room temperature with 1 mL PBST, three times for 10 min each, to remove unbound secondary antibodies (and residual dyes, if used).

3.5 Mounting

1. Remove the PBST and add sufficient anti-fade mounting medium to completely cover the samples.
2. Add a drop of the mounting medium onto a glass microscope slide and spread it with forceps to create a bed upon which the samples can be arranged on the slide.
3. Carefully transfer samples onto the glass microscope slide by picking them out of the 1.5 mL tube using forceps (*see* **Note 7**).
4. Ensure the ventral side of samples is in contact with the slide and the exposed musculature is facing up. Arrange samples according to treatment (e.g., by genotype or condition). It can be helpful to arrange samples in rows in the same orientation (Fig. 2b) (*see* **Note 8**).
5. Place cover glass gently over the samples to avoid air bubbles and seal the outside edge using nail polish.
6. Observe the samples using standard confocal fluorescence microscopy techniques (*see* **Notes 9–13**).

4 Notes

1. When using the Venus scissors to open the larval abdominal cavity, take care not to touch the fragile muscle on the ventral side.
2. Once you have added 4% PFA, proceed to dissect the next larvae. Remember to set a 40-min timer and do not fix the sample excessively.
3. Do not place more than 20 samples in a single 1.5 mL tube. Too many samples can cause damage during the washing and incubation steps.

4. Before each pipetting, place the tube vertically to sink the samples, so that the pipetting will not disrupt them. For the same reason, add the liquid against the wall of the tube rather than against the samples.
5. If the antibody is very limited, the total volume can be reduced. At the very least, however, the liquid should be able to move evenly through the tube, rather than keep at the point bottom of the tube.
6. Many diluted antibodies can be recycled, stored at 4 °C, and reused for two to three times. However, some antibodies can only be used once.
7. As samples are easily damaged by forceps, we only use forceps to hold the tail or edge of the sample, in order to make sure the part we want to examine is not touched.
8. In order to ensure that samples from the same batch share the same experimental conditions, samples of up to four genotypes can be placed in a single 1.5 mL tube for simultaneous staining. This is achieved by preserving or removing the head or tail tissue for samples of the same genotype after the fixation step, giving four distinct morphological sample types (complete sample, sample without head, sample without tail, sample without head and tail, Fig. 2b). However, only do this when you are experienced, as errors and contamination can occur easily.
9. The bouton number is commonly used as a proxy for NMJ growth [10]. To visualize and quantify the bouton number of the entire NMJ structure, the postsynaptic marker anti-discs large (anti-DLG, *Drosophila* ortholog of postsynaptic density protein 95) and presynaptic marker anti-horseradish peroxidase (anti-HRP, that specifically labels the cell membrane of neurons) are typically used. These antibodies specifically co-label only type I boutons, thereby permitting type I bouton identification and counting. As a guide, the third-instar larval NMJ typically contains 20–50 type I boutons per muscle. NMJ at muscles 6/7 has twice as many boutons because it innervates two muscles [4] and it is the most frequently examined (Fig. 3a, b). A decrease in the bouton number is considered as undergrowth [14], whereas an increase in the bouton number is considered as overgrowth [15], Sometimes, alternations in bouton number are accompanied with the bouton size change, which is measured by the area of the terminal type Ib bouton, despite the fact that they are regulated by distinct signaling pathways [16]. In addition to bouton number and size, branch point can also be used to evaluate NMJ growth. This is defined as any branch containing two or more boutons off of the primary nerve terminal as well as any subsequent branches off of these secondary branches. This is considered a measure of synaptic architecture complexity [10].

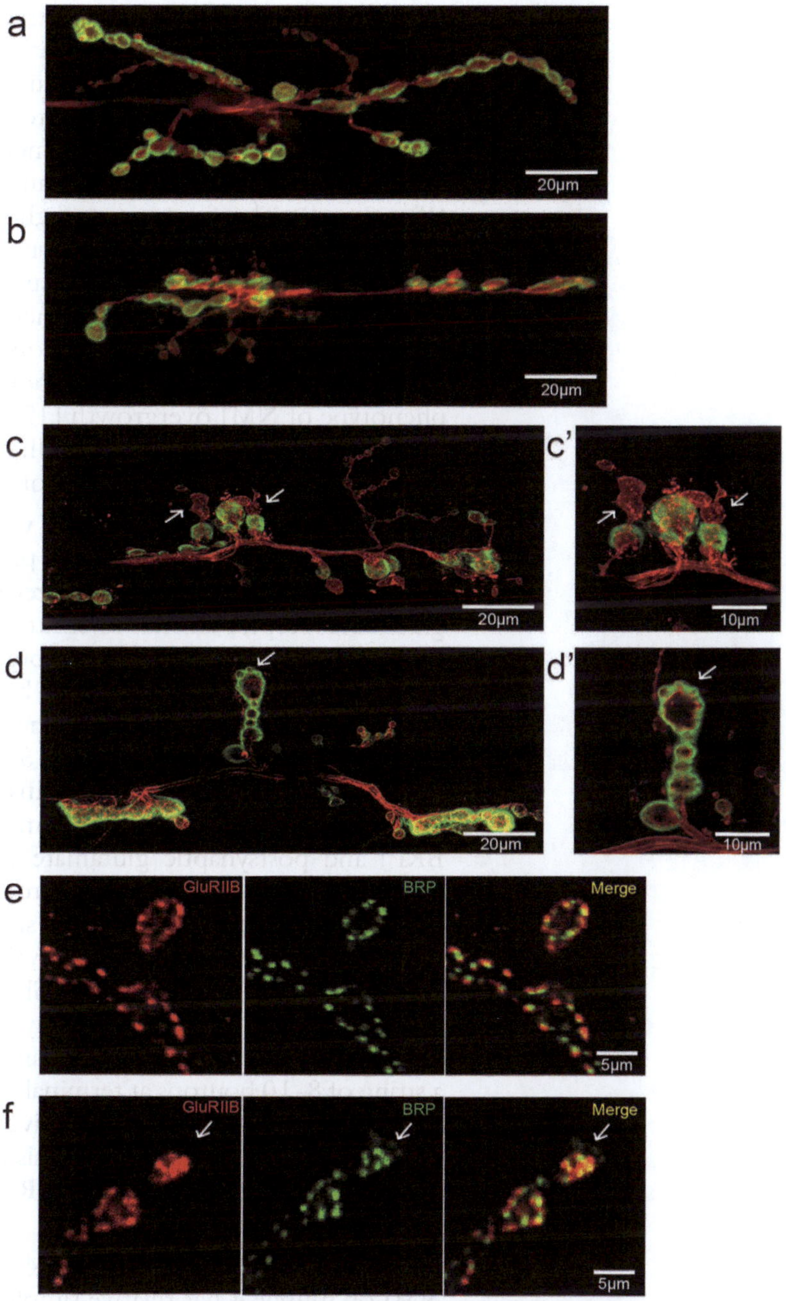

Fig. 3 Examples of NMJ phenotypes. (**a**) Wild type; the entire NMJ is shown. (**b**) A mutant NMJ with less bouton number than in wild type. (**c**) An NMJ with ghost bouton phenotype and its amplification view in **c′**. (**d**) An NMJ with satellite bouton phenotype and its amplification view in **d′**. (**e**) Wild type; the terminal boutons of an NMJ are shown, with the presynaptic BRP align precisely with the postsynaptic GluRIIB. (**f**) A mutant NMJ with unapposed active zone on bouton. The arrow head shows the BRP-positive sites without GluRIIB signal. (**a**–**d′**) are labeled with anti-DLG (green) and anti-HRP (red). (**e**, **f**) are labeled with anti-BRP and anti-GluRIIB

10. During NMJ development, many transient structures are either stabilized or retracted during the formation of the complex terminal arbor, such as ghost boutons and satellite boutons. Ghost boutons are immature boutons that contain presynaptic vesicles but lack postsynaptic elements and are therefore also known as postsynaptic footprints and orphan boutons [17, 18] (Fig. 3c, c′). The formation of ghost boutons is considered comparable to the process of synapse elimination in vertebrates; however, they form as a result of poor synaptic development [19] rather than the degeneration of mature boutons. Satellite boutons are small boutons that bud off from and surround the preexisting parent bouton and are considered a phenotype of NMJ overgrowth [17] (Fig. 3d, d′). Calculating the proportions of ghost and satellite boutons across an entire NMJ can be used as an indicator of NMJ growth.
11. Synapse stability and integrity are vital for the maintenance of mature synapse function [11]. The presynaptic neurotransmitter release sites (active zones) align precisely with the postsynaptic glutamate receptor clusters to ensure efficient synaptic transmission [20] (Fig. 3e). However, disorganization of synaptic proteins or disassembly of individual boutons might occur when synapse integrity is perturbed, a phenomenon that manifests as unapposed active zones [21] (Fig. 3f). To evaluate the arrangement of synaptic functional proteins, co-labeling is performed using the presynaptic active zone marker anti-bruchpilot (anti-BRP) and postsynaptic glutamate receptor (anti-GluR). The percentage apposed active zone area is defined as the ratio of the total overlapping BRP/GluR signal to the total BRP area (unapposed active zone area is this value subtracted from 100%). The particle number of BRP and GluR can also be measured. In the third-instar larvae, a typical type I bouton houses about 10 active zones. By counting the anti-BRP particle number on a string of 8–10 boutons at terminal of NMJ and dividing by the bouton number, the number of active zones per bouton or active zone density can be calculated. This is considered a measure of presynaptic strength. Similarly, GluR density can reflect the postsynaptic strength.
12. For subcellular imaging, muscle 4 is commonly used, as the NMJ on muscle 4 has relative simple arborization and individual bouton size is relatively large.
13. Some NMJ antigens or dyes are also strongly expressed in the muscle, such as phalloidin, which can cause a high background signal when trying to image the NMJ. In this case, analysis of muscles 12 and 13 may yield clearer images, as NMJ and muscle are more spatially separated here [22].
14. The bouton is a three-dimensional object, and thus, data from multiple optical sections (i.e., a Z-stack) should be acquired to understand the distribution of target antigens.

References

1. Ziegler AB et al (2016) The amino acid transporter JhI-21 coevolves with glutamate receptors, impacts NMJ physiology, and influences locomotor activity in drosophila larvae. Sci Rep 6. https://doi.org/10.1038/srep19692
2. Thomas U, Kobler O, Gundelfinger ED (2010) The drosophila larval neuromuscular junction as a model for scaffold complexes at glutamatergic synapses: benefits and limitations. J Neurogenet 24:109–119. https://doi.org/10.3109/01677063.2010.493589
3. Prokop A (2006) Organization of the efferent system and structure of neuromuscular junctions in Drosophila. Int Rev Neurobiol 75: 71–90. https://doi.org/10.1016/s0074-7742(06)75004-8
4. Menon KP, Carrillo RA, Zinn K (2013) Development and plasticity of the Drosophila larval neuromuscular junction. Wiley Interdiscip Rev Dev Biol 2:647–670. https://doi.org/10.1002/wdev.108
5. Marrus SB, Portman SL, Allen MJ, Moffat KG, DiAntonio A (2004) Differential localization of glutamate receptor subunits at the Drosophila neuromuscular junction. J Neurosci 24: 1406–1415. https://doi.org/10.1523/jneurosci.1575-03.2004
6. Hoang B, Chiba A (2001) Single-cell analysis of Drosophila larval neuromuscular synapses. Dev Biol 229:55–70
7. Miyazaki T et al (2021) Excitatory and inhibitory receptors utilize distinct post- and trans-synaptic mechanisms in vivo. eLife 10:e59613. https://doi.org/10.7554/eLife.59613
8. Ruiz-Cañada C, Budnik V (2006) Synaptic cytoskeleton at the neuromuscular junction. Int Rev Neurobiol 75:217–236. https://doi.org/10.1016/s0074-7742(06)75010-3
9. Chou VT, Johnson SA, Van Vactor D (2020) Synapse development and maturation at the drosophila neuromuscular junction. Neural Dev 15:11. https://doi.org/10.1186/s13064-020-00147-5
10. Miller DL, Ballard SL, Ganetzky B (2012) Analysis of synaptic growth and function in Drosophila with an extended larval stage. J Neurosci 32:13776–13786. https://doi.org/10.1523/jneurosci.0508-12.2012
11. Wagh DA et al (2006) Bruchpilot, a protein with homology to ELKS/CAST, is required for structural integrity and function of synaptic active zones in Drosophila. Neuron 49:833–844. https://doi.org/10.1016/j.neuron.2006.02.008
12. Stogsdill JA, Harwell CC, Goldman SA (2023) Astrocytes as master modulators of neural networks: synaptic functions and disease-associated dysfunction of astrocytes. Ann N Y Acad Sci. https://doi.org/10.1111/nyas.15004
13. Xing G et al (2018) Neurexin-Neuroligin 1 regulates synaptic morphology and functions via the WAVE regulatory complex in Drosophila neuromuscular junction. eLife 7. https://doi.org/10.7554/eLife.30457
14. Marqués G et al (2002) The Drosophila BMP type II receptor wishful thinking regulates neuromuscular synapse morphology and function. Neuron 33:529–543
15. Wan HI et al (2000) Highwire regulates synaptic growth in Drosophila. Neuron 26:313–329
16. Budnik V (1996) Synapse maturation and structural plasticity at Drosophila neuromuscular junctions. Curr Opin Neurobiol 6:858–867. https://doi.org/10.1016/s0959-4388(96)80038-9
17. Guangming G, Junhua G, Chenchen Z, Yang M, Wei X (2020) Neurexin and neuroligins maintain the balance of ghost and satellite boutons at the drosophila neuromuscular junction. Front Neuroanat 14:19. https://doi.org/10.3389/fnana.2020.00019
18. Eaton BA, Davis GW (2003) Synapse disassembly. Genes Dev 17:2075–2082
19. Sutcliffe B, Forero MG, Zhu B, Robinson IM, Hidalgo A (2013) Neuron-type specific functions of DNT1, DNT2 and Spz at the Drosophila neuromuscular junction. PLoS One 8: e75902. https://doi.org/10.1371/journal.pone.0075902
20. Pielage J, Fetter RD, Davis GW (2005) Presynaptic spectrin is essential for synapse stabilization. Curr Biol 15:918–928
21. Koch I et al (2008) Drosophila ankyrin 2 is required for synaptic stability. Neuron 58: 210–222
22. Ramachandran P, Barria R, Ashley J, Budnik V (2009) A critical step for postsynaptic F-actin organization: regulation of Baz/Par-3 localization by aPKC and PTEN. Dev Neurobiol 69: 583–602. https://doi.org/10.1002/dneu.20728

References

[illegible]

Chapter 17

Neural Endophenotype Assessment in Zebrafish Larvae Using Optomotor and ZebraBox Locomotion Assessment

Jiaheng Xie, Patrick Goodbourn, Tamar Sztal, and Patricia R. Jusuf

Abstract

Due to the highly conserved genetics across the central nervous system, the easily probed visual system can act as an endophenotype for assessing neurological function. Here, we describe a psychophysics approach to assess visually driven swimming behavior in the high-throughput zebrafish genetic model system. We use the optomotor response test together with general locomotion behavior to assess neural processing while excluding motor defects related to muscle function.

Key words Optomotor response function, Visual processing, Psychophysics test, Zebrafish vision, Visual swimming behavior

1 Introduction

The vertebrate zebrafish has emerged as an important model organism that fills a complementary gap in developmental biology due, in part, to its many key technical and experimental advantages. The high genetic conservation across all vertebrates combined with rapid development of zebrafish, their small size, and their ex vivo development has made this the vertebrate of choice for high-throughput studies including anatomical and behavioral phenotypes for neurodevelopmental disorders [1, 2]. For neurosciences, in which phenotypes are often complex, multifactorial, and involving integration of many cell types and circuits, the visual system affords one of the best readouts of endophenotypes due to the excellent behavioral testing ability utilizing innate responses that are easily observable. The visual system of zebrafish has thus become a great model for neural development [2–4]. Zebrafish vision becomes functional by 3 days postfertilization (dpf) [5], and visual-driven behavior is well established by 5 dpf. By the end of the first week, zebrafish deplete their yolk reserves and demonstrate visually guided feeding behavior. Here, we describe two key

Sebastian Dworkin (ed.), *Neurobiology: Methods and Protocols*, Methods in Molecular Biology, vol. 2746,
https://doi.org/10.1007/978-1-0716-3585-8_17,

experimental setups that, in combination, allow us to assess the visual endophenotypes in any condition affecting the central nervous system. These combine an optomotor swimming assay, which is driven by the innate behavior of zebrafish to swim in the same direction as a perceived visual stimulus, and the ZebraBox locomotion assay. The varying stimuli in the optomotor assay allow us to assess the range of spatial frequencies detected by the larvae and the amplitude of the response. The optomotor swimming assay is sensitive enough to assess development of the visual system across developmental times and distinguish between larvae raised with or without visual stimuli in the first 7 dpf [6, 7] based on semiautomatic quantification of displacement of zebrafish following visual stimulus presentation. The locomotion assay utilizes an automatic tracking system to quantify parameters during sustained swimming behavior in the absence of visual stimuli as a readout for general function of the central nervous system and muscle performance [8]. Quantifying the performance for both of these behavioral tests allows us to distinguish a visual neurological endophenotype from a more generalized downstream motor neuron-driven swimming phenotype. Together, these assays can compare neurological endophenotype across different genotypes (e.g., mutants generated to model human diseases or forward genetics mutagenesis screens to identify novel gene candidates), chemically induced phenotypes (e.g., toxicology or treatment screening), or different developmental stages [9, 10].

2 Materials

2.1 Optomotor Response

1. E3 medium (60×): Dissolve 34.8 g NaCl, 1.6 g KCl, 5.8 g $CaCl_2{\cdot}2H_2O$, and 9.78 g $MgCl_2{\cdot}6H_2O$ in 1.95 L distilled water. Adjust to pH 7.2 using NaOH and HCl, and adjust to final volume of 2 L with distilled water. For 1× E3 medium, dilute 16.7 mL of the 60× stock in 1 L distilled water (*see* **Note 2**).
2. Cell strainer (100 μm).
3. 90 mm Petri dish.
4. Disposable Pasteur transfer pipette.
5. Optomotor swimming arena: A number of different swimming arenas with transparent base can be designed including a linear one for zebrafish larvae (5–7 dpf). Here, we present the use of a six-lane swimming arena used to simultaneously test control (two lanes) and experimental (genetically or chemically treated in four lanes) clutches of $n = 50$ larvae per lane (Fig. 1a–c; *see* **Note 3**).
6. Power Mac G5 computer (Apple Inc., Cupertino, CA, USA) with an ATI Radeon HD 5770 graphics card (AMD, Santa Clara, CA, USA).

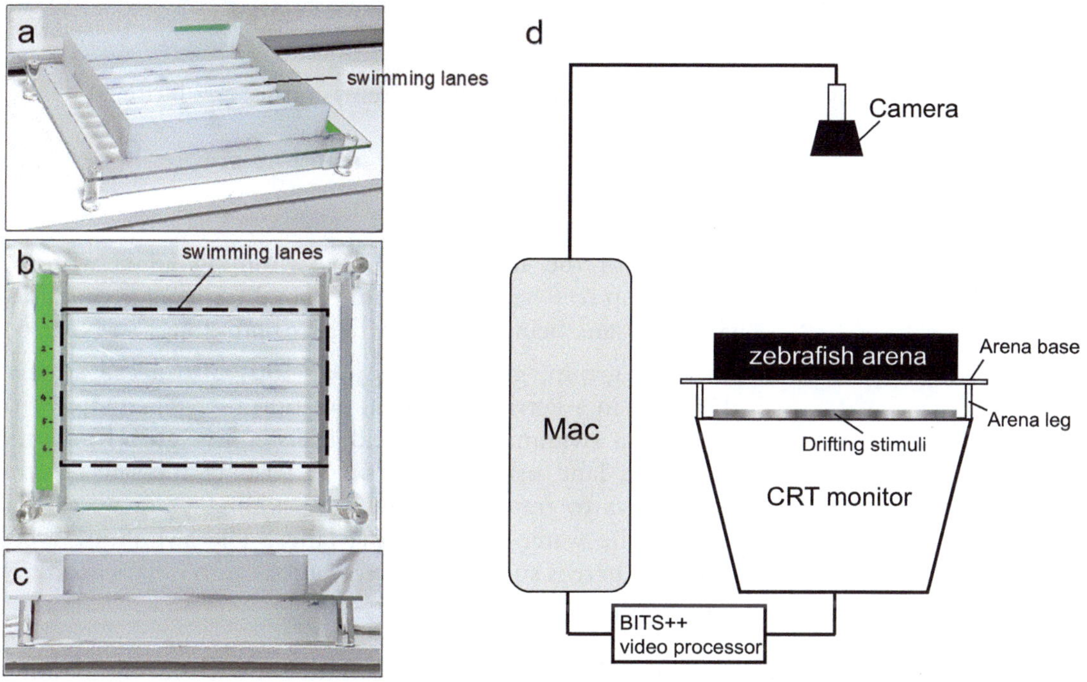

Fig. 1 Optomotor apparatus. (**a**) A picture of optomotor arena. The swimming lanes are indicated by a black dashed box from (**b**) an overhead view picture. (**c**) The front view of the arena. (**d**) Optomotor setup. Stimuli are displayed on the CRT monitor facing upward. The zebrafish swimming arena is positioned above the monitor screen, and a camera fixed above the arena obtains digital images of larvae positions before and after each test stimulus. (Adapted from [6])

7. MATLAB R2016b (MathWorks, Natick, MA, USA) with Psychtoolbox extensions [11].
8. BITS++ video processor (Cambridge Research Systems, Rochester, UK).
9. Cathode ray tube (CRT) monitor (Model M992, Dell Inc., Round Rock, TX, USA; mean luminance 50.4 cd.m^{-2}; *see* **Note 4**).
10. A C922 Pro Stream webcam (Logitech Company, Lausanne, Switzerland; *see* **Note 4**).
11. Webcam Settings software (Mactaris, Taiwan).

2.2 Locomotor Response

1. The ZebraBox system with ZebraLab software (Viewpoint, Life Sciences): This system allows automatic tracking of swimming behavior and can be used in combination with auditory and visual stimuli.
2. 24-well plastic well plates.
3. 1× E3 medium with or without 1.5% ethanol.

3 Methods

Carry out all procedures at room temperature (*see* **Note 5**).

3.1 Optomotor Response

The OMR apparatus was adapted from one previously described (Fig. 1b) [7].

3.1.1 Zebrafish Preparation

1. Fill each of the six lanes in the swimming arena with E3 medium up to a height of 5 mm (e.g., 36.3 mL for our arena with each lane being 25 mm wide, 290 mm long; *see* **Note 6**).
2. For each group, gather larvae (n ~ 50) into a cell strainer (100 μm) in a Petri dish. Then, position the cell strainer with larvae in a swimming lane, and transfer the group into the swimming lane using a disposable Pasteur transfer pipette. This allows to transfer fish without adding extra water that may lift the water depth of the lanes. The swimming arena described here is suitable for larvae aged up to 7 days postfertilization (dpf).

3.1.2 Arena Positioning

1. Carefully place the arena onto the upward-facing monitor screen without water running over swimming lanes. Use a spirit level to test whether the arena or the monitor screen is positioned evenly horizontal.
2. Use the Webcam Settings software to adjust the parameters (e.g., brightness, contrast, focus and zoom, etc.) of the webcam. Ensure that the lanes are within the scope of the camera and the fish can be clearly captured by images.
3. To block extra light from the ambient environment that may interfere the imaging, use a black curtain to cover around the sides.
4. Allow larvae to acclimatize for 10 min.

3.1.3 Stimulus Presentation and Zebrafish Imaging

Stimulus presentation can be customized depending on the endophenotype resolution required. Settings we use to assess visual performance are outlined here (Fig. 2):

1. Visual stimuli are generated by a Power Mac G5 computer running MATLAB R2016b with Psychtoolbox extensions [11] and processed on an ATI Radeon HD 5770 graphics card.
2. The outputs are sent to a BITS++ video processor for increased contrast resolution and displayed on a cathode ray tube (CRT) monitor with its screen facing upward.
3. For spatial frequency tuning function, using custom MATLAB Program, display Gaussian noise textures, filtered to achieve center spatial frequencies of 0.005, 0.01, 0.02, 0.04, 0.08, 0.16, or 0.32 c/° (SD = 0.5 octaves) below the arena (Fig. 2d) at full contrast on a CRT monitor.

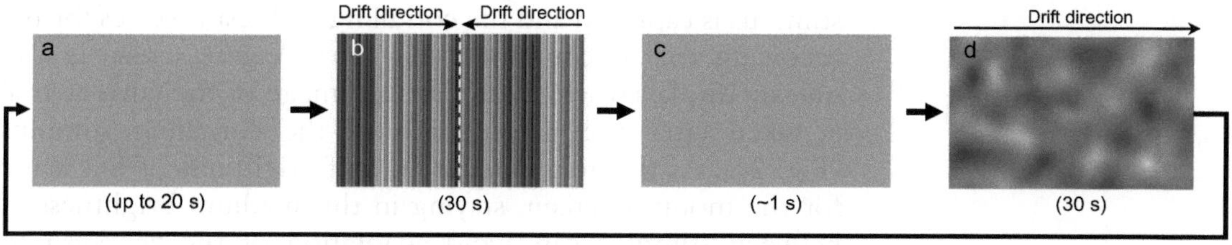

Fig. 2 An example series of stimulus display in a trial. (**a**) A gray screen during the offset of a test stimulus displayed in the previous trial. During this gray background display, a post-image of the previous trial is taken. Also, the corralling and test stimuli of the current trial are computed during this period. This display normally last up to 20 s depending on the computation time of the stimuli. Then, (**b**) larvae are corralled back to the center of the lane by high-contrast compound gratings drifting from each end for 30 s, converging at the midline (dashed white line). After the corralling stimulus, (**c**) a gray background is quickly displayed for ~1 s to allow taking a pre-image for the trial. (**d**) For test optomotor response, a test stimulus, which is a narrowband filtered Gaussian noise texture, is displayed for 30 s. This type of stimulus ensures that spatial parameters are independent of the direction the larvae are facing (unlike sinusoidal stimuli). The pattern drifts parallel to the long axis of the swimming arena lanes. After the test stimulus, the screen flashed back to a gray screen for taking a post-image for the trial. (Adapted from [6])

4. During a trial, use a range of speeds (e.g., 25, 50, or 100 °/s) to present textures drifting (at the different spatial frequency) parallel to the long axis of the lanes.
5. For contrast response function, keep the center spatial frequency at 0.02 c/° (the spatial frequency that typically gives rise to highest optomotor response; $SD = 0.5$ octaves), and present textures at 1%, 3%, 5%, 10%, 30%, 50%, or 100% contrast.
6. In a trial, present a test stimulus for 30 s with one of the combinations of parameters described. Each stimulus is presented six times (3 repeats × 2 directions) within an experimental session in a random order among other stimuli of combinations of parameters.
7. Prior to each trial, present a corralling stimulus (25 °/s drift) for 30 s to guide larvae back toward to the center of the lane. The corralling stimulus is generated using custom MATLAB program. This is a superposition of a group of random-phased sine gratings at spatial frequencies of 0.0035, 0.0071, 0.0141, 0.0283, 0.0566, 0.1131, and 0.2263 c/°, intermediate to the test frequencies at full contrast (Fig. 2b).
8. A C922 Pro Stream webcam is fixed over the arena (Fig. 1d). Use the webcam controlled by a custom MATLAB program to capture images of the whole swimming arena immediately before and after each test stimulus presentation for location of larvae in each swimming lane.
9. Display blank gray screen during the offset of the test stimulus, while the post-test image of the previously displayed test

stimulus is captured and the corralling and test textures for the following trial is computed (Fig. 2a). A gray screen is also quickly displayed while the pre-test image of the current trial is taken after the presentation of the corralling stimulus (Fig. 2c). As the gray is set to be half brightness of full white for the monitor screen, staying in this medium brightness in between stimulus can avoid adaptation of the zebrafish to either end of the brightness range (black or white) that may result in reduced sensitivity to the stimuli.

3.1.4 Optomotor Response Image Processing

1. Subtract the "before" grayscale image matrix from the "after" grayscale image matrix for a given trial. This has the effect of removing the background image (i.e., anything that has not moved) and identifying larval positions (negative values at pre-trial larval positions, positive values for posttrial larval positions).

Box 1 Example MATLAB Code for Extracting Larval Positions

```
% Convert preimage and postimage into variables; 'pre-
image' and 'postimage' here are the file names of images
taken before and after a trial, respectively.
PreIm = double(imadjust(imread(preimage)));
PstIm = double(imadjust(imread(postimage)));
%Extract larval positions
diffIm = PreIm - PstIm;
```

2. In order to identify the larvae, the contrast of the subtracted image is flattened, and a Laplacian-of-Gaussian filter was applied to highlight edges [12]. Larvae need to be segmented by thresholding (manually adjusted if necessary).
3. Calculate the centroid of the larval positions using *bwconncomp* and *regionprops* commands from the MATLAB Image Processing Toolbox.

3.1.5 Optomotor Index Calculations

1. For each trial, compute the optomotor index (OMI) using the change in position of the group centroid (i.e., average position of the group) in the direction of the texture motion.
2. For spatial frequency tuning functions, normalize OMI by normalizing all data to the highest averaged OMI of wild-type larvae across tested combinations of parameters described. In our data, the highest averaged OMI was typically at 0.02 c/° and 25 °/s.

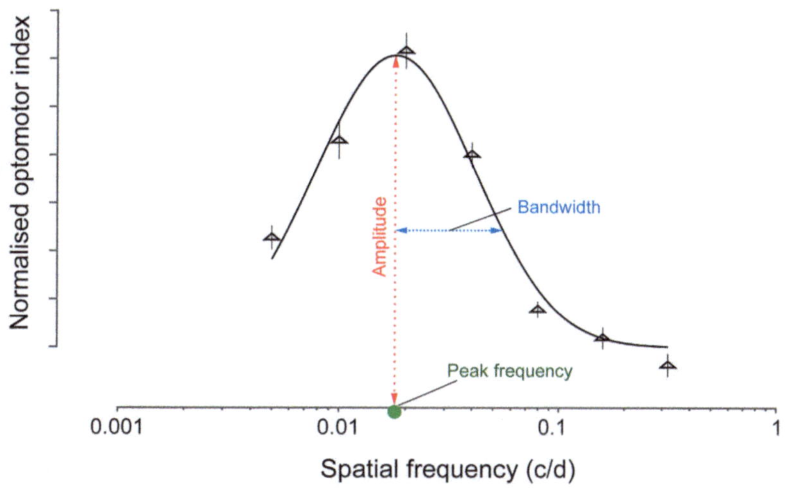

Fig. 3 Sample spatial frequency tuning function from larval zebrafish optomotor response (OMR). The black curve represents a three-parameter log-Gaussian function, which provides an excellent fit to the data. The three free parameters of the fit are depicted: amplitude (i.e., height of the peak) is shown as a red dashed line; bandwidth (i.e., standard deviation) is shown as a blue dashed line; and peak frequency (i.e., the spatial frequency at peak amplitude) is shown as a green point on the x axis

3. For contrast response functions, normalize OMIs to the highest averaged OMI of the wild-type group across tested combinations of speeds and contrasts at 0.02 c/°. Theoretically, the highest averaged should be at 100% contrast at either 25 or 50 °/s of speed.

3.1.6 Statistical Analysis of Optomotor Index

1. Fit the spatial frequency tuning function as a log-Gaussian using a least squares criterion. From each fitted model, if required, compute estimated amplitude (i.e., height of the peak), peak spatial frequency (i.e., spatial frequency at which amplitude peaked), and bandwidth (i.e., standard deviation; Fig. 3).
2. Fit the contrast response function as a two-parameter piecewise function (Eq. 1) using a least squares criterion:

$$y = \begin{cases} 0, & (x < k) \\ (x - k)^a, & (x \geq k) \end{cases} \tag{1}$$

where y is the OMI, x is log stimulus contrast, k is the contrast threshold (i.e., minimum contrast to evoke an optomotor response), and a is response gain (i.e., slope of the function; Fig. 4; *see* **Note 8**).
3. To test whether spatial frequency tuning and contrast response functions differ between groups, use an omnibus F test to

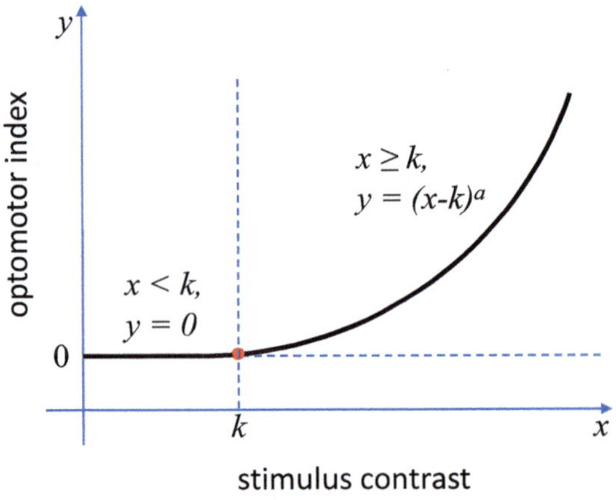

Fig. 4 A schematic of the contrast response function. The black curve represents a two-parameter piecewise function, which provided an excellent fit to our data. The optomotor response arises after the contrast threshold (*k*, red dots on the *x* axis dimension) and increases as an exponential-like pattern with a slope *a* (i.e., response gain)

compare the goodness of fit (r^2) of a full model, in which parameter estimates of each group could vary independently, with that of a restricted model, in which parameters were constrained to be the same across groups.

4. To determine whether specific parameter estimates differ between groups, use nested *F* test to compare a full model with a restricted model in which one parameter was constrained to be the same across groups. A criterion of $\alpha = 0.05$ was used to determine significance, with Bonferroni correction applied to P values to account for multiple testing where appropriate [13] (*see* **Note 9**).

3.2 ZebraBox Locomotor Response

3.2.1 Experimental Locomotion Assay Setup

1. The assay can be set up in E3 embryo medium or with various activators of swimming, for example, 1.5% ethanol. Pretreat zebrafish with testing solution in a Petri dish for 10 min prior to assaying.
2. Transfer larvae into 24-well dishes, containing 1 mL of solution, in a pseudorandom order, with the experimenter blinded to experimental conditions (e.g., genotype or treated individuals) during testing and data analysis.
3. Place well plates into the recording chamber and ensure all wells are visible inside the circular grid on the software in order to detect each larva.
4. Using the software "generate new protocol" and set duration and integration time in "protocol parameters." Set the recording areas by highlighting the entire grid, and define the location of a well.

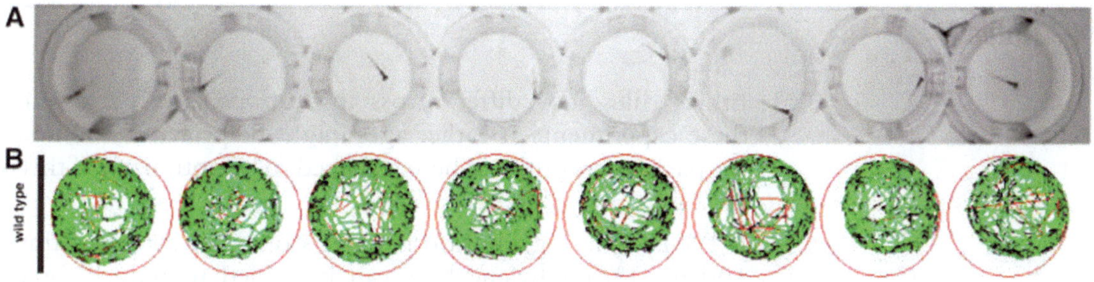

Fig. 5 Locomotor response. (**a**) Individual zebrafish embryos are placed into wells in a 48-well plate. (**b**) Locomotion is recorded using an infrared digital camera. The zebrafish movements during the entire test period can be visualized and are color coded based on the detection thresholds entered in the software (red, fast movement; green, slow movements; black, inactivity). (Adapted from [8])

5. For 7 dpf WT larvae, set an inactivity threshold of 1 mm/s, a detection threshold of 25 mm/s, and a maximum burst threshold of 30 mm/s.
6. Zebrafish were allowed to swim for 10 min in the dark (0% setting under "light driving settings").
7. Close the chamber and start the video recording. The video is captured using an infrared light source (Fig. 5).

3.2.2 ZebraBox Assay Analysis

1. After completion of the experiment, three files will be generated. An Excel sheet with parameters including small vs. large movement counts and distance covered by fish in small vs. large movements. Additionally, a video file of the recording period and an image file containing a visual representation of the locomotion are provided.
2. For the 10-min test period in the dark, the total distance swum and the swimming speed above inactivity threshold and below maximum burst threshold are extracted using ZebraLab software (Viewpoint Life Sciences).
3. For each trial, test for indicators of normality using D'Agnostino and Pearson's test [14]; and identify and exclude outliers using the ROUT method.
4. Depending on the experimental design and question, different statistical tests should be used. For our comparison of experimental groups, use two-way ANOVA (genotype × treatment) with Bonferroni correction in Prism 7 (GraphPad, San Diego, CA, USA) ($\alpha = 0.05$). For robust results, at least $n = 100$ larvae per group should be used for analysis of swimming distance and speed for each combination of experimental control and ZebraBox assay (e.g., dark only, light introduced during the testing period, with and without ethanol).

4 Notes

1. Ethics rules may differ across institutions, and performing these experiments at earlier ages might be preferred. Some of the parameters outlined here will need to be optimized, if this is the case.
2. If E3 1× working solution is to be stored for extended time, 100 μL 1% methylene blue can be added to reduce fungal growth.
3. Due to the limit of the screen resolution, the base of our swimming arena was designed to be lifted from the screen, so that a higher limit of spatial frequency for test can be achieved. Displaying a spatial frequency of stimulus with a shorter viewing distance results in finer texture. Strange patterns may be present when the fineness of the texture exceeds the screen resolution. However, it has been noted that visual stimulus distortion frequently occurs for behavior experiments using aquatic species [15]; the light bends at the interface between water and air as well as the media in between the two (e.g., plastic), so the actual visual stimulus the fish perceive may not be well as what researchers intend. Our design is sensitive for comparing visual function between fish with different genetic backgrounds (wild type vs. mutants) or under different treatments (treated or untreated with a drug) but may not suitable for zebrafish visual characterization, such as spatial frequency limit or visual acuity. To minimize this distortion for such goals, it is important to have as higher ratio of zebrafish's view distance in the water to the distance of other media between the swimming lanes and the stimulus display as possible.
4. The OMR method presented here was designed for displaying stimuli using a CRT monitor, but the setup can be easily adapted to utilize a LCD monitor, which is currently the main commercial type. LCD monitor has already been used for zebrafish OMR analysis [16]. As this setup takes images only before and after each test stimulus, a common commercially available webcam, which normally comes with a framerate of 30–60 frames per second, will fulfill the need for the recording. Required camera resolution is dependent on imaging distance between the camera and the swimming lanes. For our setup, we use a webcam with a resolution of 1920 × 1080 pixels for an imaging distance of 366.5 mm.
5. Consistency in temperature is also important. As the ideal temperature for zebrafish is 28.5 °C, both tests should be run ideally close to that temperature, if possible, in temperature-controlled rooms or at least >24 °C.

6. The OMR swimming arena used must be designed to ensure enough E3 volume for the chosen age and density of the zebrafish (usually defined in the relevant ethics approval). Additionally, the height of the E3 should be kept at a minimum (our swimming lane uses 5 mm height), so that the zebrafish are relatively equidistant from the stimulus, which will affect the spatial frequency seen by the fish.
7. Consistency of timing is important, and the test and control groups must be assessed together. Visual development at these ages progresses rapidly, so comparison between batches performed in the morning vs. late afternoon will introduce unwanted variations. We recommend two control and four treatment (e.g., morpholino injected, mutant, or chemically exposed larvae) lanes in each experimental run. We recommend at least duplicates for this test and ensure at least four groups of control fish tested for a narrow data deviation.
8. The two-parameter piecewise function for contrast response analysis shown here is based on the fact that the data we got previously did not present saturation before the higher limit of contrast. An experimental design with a higher contrast resolution, especially in the higher range, may detect a peak response before the full contrast and show a plateau in the higher end. In this case, a four- or five-parameter sigmoidal model will provide a better fit to the data.
9. Statistics for comparing the fit of models or parameters between OMR datasets is based on model selection. Detailed descriptions of this method can be found in Lu and Dosher, MIT Press, 2013 [13].
10. Zebrafish activity changes throughout the day, for consistency locomotion assays should be performed at the same time, ideally between 9 am and 12 pm, when larvae are most active. Due to variation in the response, we recommend at least six replicates. Performing a power calculation based on pilot data is important, to confidently identify or rule out differences between strains or conditions.
11. More detailed description of software menu options for the ZebraBox are described in Sztal et al., 2016 JOVE [8].

References

1. Curpan AS, Balmus IM, Dobrin RP et al (2022) A mini-review regarding the modalities to study neurodevelopmental disorders-like impairments in zebrafish-focussing on neurobehavioural and psychological responses. Brain Sci 12(9). https://doi.org/10.3390/brainsci12091147
2. Bilotta J, Saszik S (2001) The zebrafish as a model visual system. Int J Dev Neurosci 19(7):621–629. https://doi.org/10.1016/s0736-5748(01)00050-8
3. Chhetri J, Jacobson G, Gueven N (2014) Zebrafish--on the move towards ophthalmological

research. Eye (Lond) 28(4):367–380. https://doi.org/10.1038/eye.2014.19
4. Gestri G, Link BA, Neuhauss SC (2012) The visual system of zebrafish and its use to model human ocular diseases. Dev Neurobiol 72(3): 302–327. https://doi.org/10.1002/dneu.20919
5. Easter SS Jr, Malicki JJ (2002) The zebrafish eye: developmental and genetic analysis. Results Probl Cell Differ 40:346–370. https://doi.org/10.1007/978-3-540-46041-1_17
6. Xie J, Jusuf PR, Bui BV et al (2019) Experience-dependent development of visual sensitivity in larval zebrafish. Sci Rep 9(1): 18931. https://doi.org/10.1038/s41598-019-54958-6
7. Orger MB, Gahtan E, Muto A et al (2004) Behavioral screening assays in zebrafish. Methods Cell Biol 77:53–68. https://doi.org/10.1016/s0091-679x(04)77003-x
8. Sztal TE, Ruparelia AA, Williams C et al (2016) Using touch-evoked response and locomotion assays to assess muscle performance and function in zebrafish. J Vis Exp 116. https://doi.org/10.3791/54431
9. Neuhauss SC (2003) Behavioral genetic approaches to visual system development and function in zebrafish. J Neurobiol 54(1): 148–160. https://doi.org/10.1002/neu.10165
10. Neuhauss SC, Biehlmaier O, Seeliger MW et al (1999) Genetic disorders of vision revealed by a behavioral screen of 400 essential loci in zebrafish. J Neurosci 19(19):8603–8615. https://doi.org/10.1523/JNEUROSCI.19-19-08603.1999
11. Mario K, David B, Denis P et al (2007) What's new in Psychtoolbox-3. Perception 36(14): 1–1
12. Bizrah M, Dakin SC, Guo L et al (2014) A semi-automated technique for labeling and counting of apoptosing retinal cells. BMC Bioinf 15(1):1–12
13. Lu Z-L, Dosher B (2013) Visual psychophysics: from laboratory to theory. MIT Press
14. D'AGOSTINO R, Pearson ES (1973) Tests for departure from normality. Empirical results for the distributions of b 2 and√ b. Biometrika 60(3):613–622
15. Dunn TW, Fitzgerald JE (2020) Correcting for physical distortions in visual stimuli improves reproducibility in zebrafish neuroscience. eLife 9:e53684. https://doi.org/10.7554/eLife.53684
16. Creton R (2009) Automated analysis of behavior in zebrafish larvae. Behav Brain Res 203(1): 127–136. https://doi.org/10.1016/j.bbr.2009.04.030

Chapter 18

The Photothrombotic Model of Ischemic Stroke

David E. Wong Zhang, Shenpeng R. Zhang, Hyun Ah. Kim, Christopher G. Sobey, and T. Michael De Silva

Abstract

Stroke is a major cause of morbidity worldwide; yet, there is a lack of treatment options to address post-stroke cognitive and motor impairment, thus there is an urgency for developing neuroprotective and restorative therapies. Much of our fundamental understanding of stroke pathology has been derived from animal models. The photothrombotic model of ischemic stroke is commonly used to study cellular and molecular mechanisms of neurodegeneration, test functional/cognitive outcomes, identify important biomarkers, and assess the effectiveness of novel therapies. It allows for the precise targeting of an infarct to a specific region of the brain, has a low mortality rate, low seizure rate, and is relatively easy to perform. This chapter outlines materials and methods for the photothrombotic model of ischemic stroke in mice, its limitations, and some considerations needed when using this model.

Key words Brain ischemia, Rodents, Prefrontal cortex, Cognition, Motor cortex, Motor impairment

1 Introduction

Stroke is the second most common cause of mortality after ischemic heart disease and a major cause of morbidity worldwide [1]. The lifetime risk of having a stroke is estimated at 25%, irrespective of sex [2]. Without improved risk factor management or development of new therapeutics, the burden of stroke is expected to rise [1]. Stroke is caused by the occlusion and/or rupture of a cerebral blood vessel, leading to tissue hypoxia. Recent reports suggest that globally 62% of strokes are ischemic (caused by artery occlusion by an embolus or thrombus), while 38% are hemorrhagic (caused by either intracerebral or subarachnoid bleeding) [1]. In ischemic stroke, a series of pathological events occurs after the initial occlusion (e.g., excitotoxicity, oxidative stress, mitochondrial dysfunction, and inflammation) that contribute to the formation of an ischemic core [3]. The surrounding partially perfused tissue known as the penumbra survives but is inactive. Unless blood

Sebastian Dworkin (ed.), *Neurobiology: Methods and Protocols*, Methods in Molecular Biology, vol. 2746,
https://doi.org/10.1007/978-1-0716-3585-8_18,

flow is restored, the cells within the penumbra will also eventually die, leading to an expansion of the infarct. Death of neurons within the infarct may disrupt neural circuits, which can contribute to secondary neurodegeneration and loss of function [4]. As the period of ischemia is extended, the outcomes are typically worsened [5]. Therefore, it is paramount for early intervention once symptoms of stroke manifest, as early recanalization and restoration of cerebral blood flow lead to better outcomes [6, 7].

Current treatments for stroke are limited to reperfusion therapies, namely, intravenous recombinant tissue plasminogen activator (r-tPA) and mechanical thrombectomy. However, relatively few stroke patients are eligible to receive r-tPA as it has a short therapeutic window from stroke onset (generally within 4.5 h) [8]. Mechanical thrombectomy is only suitable for clots located in larger cerebral arteries, and has a diminishing benefit of outcome over time [9]. Furthermore, numerous contraindications such as uncontrolled hypertension, stroke history, coagulopathy, and hyper- or hypoglycemia further limit eligibility for thrombectomy [10]. Importantly, neurological recovery following successful recanalization is not guaranteed [11], highlighting the need for new therapies that address more than mere restoration of blood supply.

Research utilizing animal models has advanced our understanding of stroke pathophysiology [3], identifying therapeutic targets to reduce infarct volume [12, 13], infection [14], and neurological deficits [15], thus helping to inform decisions to improve stroke outcomes in patients. However, there is still a considerable need for further research due to a lack of therapies for patients who are ineligible for r-tPA or thrombectomy. A common method used in ischemic stroke research involves an induction of photothrombosis of the cerebral cortical vessels in rodents. First developed in 1985 [16], the photothrombotic model of stroke is relatively simple to use when compared to the middle cerebral artery occlusion model and allows for the specific targeting of a region of the brain cortex with high reproducibility [17, 18]. It involves the systemic (intraperitoneal) administration of a photosensitive dye which is then activated within the circulation by a light source through the intact skull of a mouse. Studies have commonly used Rose Bengal as the photosensitive dye of choice as its mechanism of occlusion resembles that of atherothrombosis [19]. Other photosensitive dyes include Evans blue and erythrosine B [17, 20, 21]. The photothrombotic model is a valuable technique for evaluating cortical plasticity [22], secondary neurodegeneration [23], efficacy of therapies designed to promote regeneration [24], and cognitive/functional rescue [25]. In this chapter, we outline procedures for performing the photothrombotic model of ischemic stroke in mice.

2 Materials

2.1 Photosensitive Dye Preparation

Prepare a Rose Bengal solution (10 mg/mL) in 0.9% sodium chloride (NaCl). Each mouse (25–35 g) will require 0.2 mL of 10 mg/mL Rose Bengal solution.

Rose Bengal

Injectable saline (0.9% NaCl)

Microcentrifuge tube

2.2 Surgical Instruments and Setup

Surgical gown

Sterile surgical gloves

Anesthetic machine—inhalation (isoflurane) ×2 (e.g., Darvall DVM-Iso Vaporizer)

1. Oxygen tank
2. Oxygen flow meter
3. Gas scavenging canister
4. Heated nose cone
5. Induction chamber

Local anesthetic—bupivacaine 2.5 mg/kg (*see* **Note 1**)

Analgesic—carprofen 5 mg/kg (*see* **Note 1**)

Heat mat

Surgical drape

Polyvisc® eye ointment (containing paraffin 30 mg/g and wool fat 30 mg/g)

Animal hair clipper

4% (w/v) chlorhexidine-s scrub

0.5% (w/v) chlorhexidine-c in 70% ethanol

Sterile cotton tip applicators

Cotton tip (non-sterile)

Gauze

Vetbond™ tissue adhesive (3 M Animal Care Products)

Sterile #15 scalpel blades and a handle

Forceps (Fine Science Tools Standard Pattern Forceps 11000-16; *see* **Note 2**)

Standard U-frame stereotaxic instrument (Model 900LS Small Animal Stereotaxic, David Kopf instruments)

1. Stereotaxic frame nose cone mask
2. 50 mm non-rupture ear bars

3. 20× objective lens
4. LED Cold Light Source KL 1600 with gooseneck (KL 1600 LED, Schott)
5. Digital display console (Model 940, David Kopf Instruments)

Homeothermic monitoring system (Harvard apparatus)

Digital timer

1 mL syringes and 27-gauge needles

Superfine marker

3 Methods

All appropriate PPE should be worn; surgical instruments and workspaces should be sterilized prior to commencing surgical procedures to reduce the risk of infection in animals. Animals are to be under anesthesia throughout the entirety of this procedure.

3.1 Anesthesia and Surgical Preparation

1. Prepare anesthetic and analgesic in accordance with animal weight.
2. Place the animal in the induction chamber filled with 5% isoflurane in 100% O_2 at a delivery rate of 5 L/min.
3. Once anesthetized, place the animal on a heat mat with a heated nose cone for continuous delivery of isoflurane at 2% in 100% O_2 at 0.4–0.6 L/min.
4. Apply eye ointment on both eyes to prevent damage to the cornea.
5. Inject the animal with analgesic on either side of the flank subcutaneously.
6. Use an animal hair clipper to shave the head, starting from the base of the skull to the nasal region.
7. Cover the animal with surgical drape.
8. With sterile cotton tips, apply the chlorhexidine-s scrub three times followed by a single application of chlorhexidine-c scrub.
9. Inject local anesthetic on the head where the incision is to be made.
10. Conduct a hind paw withdrawal test to check the depth of anesthesia. If no withdrawal reflex is observed, proceed to making an incision.

3.2 Surgery and Stereotaxic Instrument Setup

1. Using a scalpel blade, make a 2 cm midline incision on the head to expose the skull.
2. Use a sterile cotton tip applicator to remove any blood and/or connective tissue.

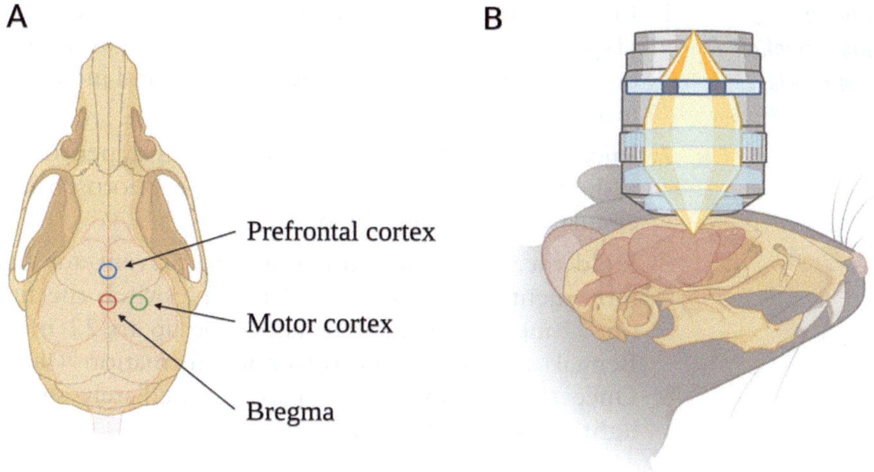

Fig. 1 Methodological considerations for photothrombotic stroke. (**a**) The bony landmark bregma lies at the intersection between the sagittal and coronal sutures (indicated by the red circle); the prefrontal cortex is located 1.2 mm anterior to the bregma (indicated by the blue circle); the M1 primary motor cortex is located 1.5 mm laterally from the bregma (either left or right, indicated by the green circles). (**b**) It is critical to elevate the nose point to make the top surface of the skull parallel to the objective lens. Created with Biorender.com

3. Once the skull has dried, identify the bony landmark bregma, and mark with a superfine marker (Fig. 1).
4. Transfer the animal to the stereotaxic apparatus.
5. Fit the animal's teeth over the incisor bar. Cover the nose with the stereotaxic frame nose cone to deliver 2% isoflurane in 100% O_2 at 0.4–0.6 L/min.
6. Fit the ear bars into the animal's ear canal to secure and center the head. Adjust the height of the head so that the curvature between the bregma and targeted coordinates aligns with the transverse plane as closely as possible.
7. Insert the rectal probe to monitor and maintain core body temperature at 37 ± 0.5 °C.
8. Switch on the light source to the lowest visible setting.
9. Lower the objective lens to target the bregma. Reset X, Y, and Z coordinates on the digital display console to 0. Reposition the light source using the digital console to precisely target the region of interest (*see* **Note 3**). Further lower the objective lens until it is in contact with the skull but avoid excessive pressure and light scattering (*see* **Note 4**).
10. Switch off the light source and set illumination to the highest setting (680 lm).

3.3 Rose Bengal Injection and Induction of Photothrombosis

1. Ensure that the animal's core body temperature is 37 ± 0.5 °C. Hypo- or hyperthermia during the induction of photothrombosis may result in infarct size variability and increase the risk of mortality.
2. Inject 0.2 mL of Rose Bengal solution intraperitoneally. Using a digital timer, allow 5 min for the photosensitive dye to circulate throughout the body.
3. Hold the LED cold light source to prevent any unnecessary movement of the objective lens and switch the light source on. Using a digital timer, allow for up to 22 min for Rose Bengal activation and thrombus formation (*see* **Note 5**). Once the set time has elapsed, immediately switch off the light source (*see* **Note 6**).

3.4 Wound Closure and Recovery

1. Remove the animal from the stereotaxic apparatus and place on a heat mat with a heated nose cone maintaining 2% isoflurane in 100% O_2 at 0.4–0.6 L/min.
2. Use a sterile cotton tip applicator or sterile forceps to bring the skin of the wound together, apply Vetbond™ tissue adhesive.
3. Cease anesthesia delivery but maintain O_2 supply. Once the animal has regained consciousness, place them in a recovery box half exposed to a pre-warmed heat mat. Provide recovery food ad libitum. Observe regularly (every 15 min) until the animal is fully awake (*see* **Note 7**). Provide analgesic once daily for the first 3 days (*see* **Note 8**).

3.5 Limitations and Considerations

The photothrombotic model typically results in permanent focal ischemia to the cortical region of the brain. The infarct may extend to the subcortical region with a longer period of illumination [17]. Infarct size may be adjusted by changing the duration of the light exposure [17, 22]. The procedure is minimally invasive and has a low mortality rate (<5%) using the methodology described above. However, patients suffering from ischemic stroke often present with a single transient LVO [26] rather than a regional permanent microvascular occlusion. Furthermore, spontaneous reperfusion injuries are a common feature of stroke [27] which does not occur with the procedure described above [28]. To address this, modifications to the protocol as described by Clark et al. may be needed to induce reperfusion post-occlusion [28]. This involves the creation of a cranial window, the use of a green diode laser for photothrombosis, a digital micro-mirror device for precise targeting and laser speckle contrast imagining, and finally, an infrared laser to induce recanalization of the occluded artery. However, the reperfusion protocol can only be used when targeting the middle cerebral artery. Furthermore, this photothrombotic model causes vasogenic edema which is thought to compromise the development of a penumbra [21]. As the penumbra is a major target of neuroprotective therapies in acute ischemic stroke [29, 30], this protocol may not be suitable for such studies. Another consideration is that the photothrombotic model is unresponsive to r-tPA treatment [31], as the thrombus formed lacks the fibrin content typically seen in ischemic strokes [32]. Recent studies have implemented a mixture of thrombin and Rose Bengal, creating r-tPA sensitive clots, to circumvent this issue [33, 34]. However, the use of thrombin in this model may introduce inflammatory factors that can influence disease progression and outcome [35]. Lastly, differences in animal strain may result in differences in functional and morphological outcomes [17]. Thus, careful considerations are needed when designing and conducting experiments using this model.

4 Notes

1. Choice of anesthesia and analgesia should be determined by the user in accordance with institutional and other regulatory guidelines.
2. Suggested surgical instruments. Similar instruments may be suitable.
3. Prefrontal cortex is targeted 1.2 mm anterior from the bregma; the primary motor cortex is targeted 1.5 mm laterally from the bregma (this may also target a part of the somatosensory region of the mouse brain; Fig. 1a) [36].
4. Light scattering indicates uneven contact between the skull and objective lens which can result in infarct size variability. To avoid, adjust the angle of the skull by elevating or lowering the head on the stereotaxic apparatus; ensure the objective lens makes even contact with the skull (Fig. 1b).
5. Infarct size can be adjusted with the illumination time [17]. Common durations for this model are either 15, 18, 20, or 22 min [17, 25].
6. Occlusion of the cerebral blood vessels can be verified via a laser speckle contrast imaging of the exposed skull after photothrombosis.
7. Postsurgical monitoring should comply with institutional and other regulatory guidelines.
8. Postsurgical administration of analgesics should comply with institutional and other regulatory guidelines.

Acknowledgments

Figure created with BioRender.com

References

1. Feigin VL, Stark BA, Johnson CO, Roth GA, Bisignano C, Abady GG, Abbasifard M, Abbasi-Kangevari M, Abd-Allah F, Abedi V, Abualhasan A, Abu-Rmeileh NM, Abushouk AI, Adebayo OM, Agarwal G, Agasthi P, Ahinkorah BO, Ahmad S, Ahmadi S, Ahmed Salih Y, Aji B, Akbarpour S, Akinyemi RO, Al Hamad H, Alahdab F, Alif SM, Alipour V, Aljunid SM, Almustanyir S, Al-Raddadi RM, Al-Shahi Salman R, Alvis-Guzman N, Ancuceanu R, Anderlini D, Anderson JA, Ansar A, Antonazzo IC, Arabloo J, Ärnlöv J, Artanti KD, Aryan Z, Asgari S, Ashraf T, Athar M, Atreya A, Ausloos M, Baig AA, Baltatu OC, Banach M, Barboza MA, Barker-Collo SL, Bärnighausen TW, Barone MTU, Basu S, Bazmandegan G, Beghi E, Beheshti M, Béjot Y, Bell AW, Bennett DA, Bensenor IM, Bezabhe WM, Bezabih YM, Bhagavathula AS, Bhardwaj P, Bhattacharyya K, Bijani A, Bikbov B, Birhanu MM, Boloor A, Bonny A, Brauer M, Brenner H, Bryazka D, Butt ZA, Caetano dos Santos FL, Campos-Nonato IR, Cantu-Brito-C, Carrero JJ, Castañeda-Orjuela CA, Catapano AL, Chakraborty PA, Charan J, Choudhari SG, Chowdhury EK, Chu D-T, Chung S-C, Colozza D, Costa VM, Costanzo S, Criqui MH, Dadras O, Dagnew B, Dai X, Dalal K, Damasceno AAM, D'Amico E, Dandona L, Dandona R, Darega Gela J, Davletov K, De la Cruz-Góngora V, Desai R, Dhamnetiya D, Dharmaratne SD, Dhimal ML, Dhimal M, Diaz D, Dichgans M, Dokova K, Doshi R, Douiri A, Duncan BB, Eftekharzadeh S, Ekholuenetale M, El Nahas N, Elgendy IY, Elhadi M, El-Jaafary SI, Endres M, Endries AY, Erku DA, Faraon EJA, Farooque U, Farzadfar F, Feroze AH, Filip I, Fischer F, Flood D, Gad MM, Gaidhane S, Ghanei Gheshlagh R, Ghashghaee A, Ghith N, Ghozali G, Ghozy S, Gialluisi A, Giampaoli S, Gilani SA, Gill PS, Gnedovskaya EV, Golechha M, Goulart AC, Guo Y, Gupta R, Gupta VB, Gupta VK, Gyanwali P, Hafezi-Nejad N, Hamidi S, Hanif A, Hankey GJ, Hargono A, Hashi A, Hassan TS, Hassen HY, Havmoeller RJ, Hay SI, Hayat K, Hegazy MI, Herteliu C, Holla R, Hostiuc S, Househ M, Huang J, Humayun A, Hwang B-F, Iacoviello L, Iavicoli I, Ibitoye SE, Ilesanmi OS, Ilic IM, Ilic MD, Iqbal U, Irvani SSN, Islam SMS, Ismail NE, Iso H, Isola G, Iwagami M, Jacob L, Jain V, Jang S-I, Jayapal SK, Jayaram S, Jayawardena R, Jeemon P, Jha RP, Johnson WD, Jonas JB, Joseph N, Jozwiak JJ, Jürisson M, Kalani R, Kalhor R, Kalkonde Y, Kamath A, Kamiab Z, Kanchan T, Kandel H, Karch A, Katoto PD, Kayode GA, Keshavarz P, Khader YS, Khan EA, Khan IA, Khan M, Khan MA, Khatib MN, Khubchandani J, Kim GR, Kim MS, Kim YJ, Kisa A, Kisa S, Kivimäki M, Kolte D, Koolivand A, Koulmane Laxminarayana SL, Koyanagi A, Krishan K, Krishnamoorthy V, Krishnamurthi RV, Kumar GA, Kusuma D, La Vecchia C, Lacey B, Lak HM, Lallukka T, Lasrado S, Lavados PM, Leonardi M, Li B, Li S, Lin H, Lin R-T, Liu X, Lo WD, Lorkowski S, Lucchetti G, Lutzky Saute R, Magdy Abd El Razek H, Magnani FG, Mahajan PB, Majeed A, Makki A, Malekzadeh R, Malik AA, Manafi N, Mansournia MA, Mantovani LG, Martini S, Mazzaglia G, Mehndiratta MM, Menezes RG, Meretoja A, Mersha AG, Miao Jonasson J, Miazgowski B, Miazgowski T, Michalek IM, Mirrakhimov EM, Mohammad Y, Mohammadian-Hafshejani A, Mohammed S, Mokdad AH, Mokhayeri Y, Molokhia M, Moni MA, Montasir AA, Moradzadeh R, Morawska L, Morze J, Muruet W, Musa KI, Nagarajan AJ, Naghavi M, Narasimha Swamy S, Nascimento BR, Negoi RI, Neupane Kandel S, Nguyen TH, Norrving B, Noubiap JJ, Nwatah VE, Oancea B, Odukoya OO, Olagunju AT, Orru H, Owolabi MO, Padubidri JR, Pana A, Parekh T, Park E-C, Pashazadeh Kan F, Pathak M, Peres MFP, Perianayagam A, Pham T-M, Piradov MA, Podder V, Polinder S, Postma MJ, Pourshams A, Radfar A, Rafiei A, Raggi A, Rahim F, Rahimi-Movaghar V, Rahman M, Rahman MA, Rahmani AM, Rajai N, Ranasinghe P, Rao CR, Rao SJ, Rathi P, Rawaf DL, Rawaf S, Reitsma MB, Renjith V, Renzaho AMN, Rezapour A, Rodriguez JAB, Roever L, Romoli M, Rynkiewicz A, Sacco S, Sadeghi M, Saeedi Moghaddam S, Sahebkar A, Saif-Ur-Rahman K, Salah R, Samaei M, Samy AM, Santos IS, Santric-Milicevic MM, Sarrafzadegan N, Sathian B, Sattin D, Schiavolin S, Schlaich MP, Schmidt MI, Schutte AE, Sepanlou SG, Seylani A, Sha F, Shahabi S, Shaikh MA, Shannawaz M, Shawon MSR, Sheikh A, Sheikhbahaei S,

Shibuya K, Siabani S, Silva DAS, Singh JA, Singh JK, Skryabin VY, Skryabina AA, Sobaih BH, Stortecky S, Stranges S, Tadesse EG, Tarigan IU, Temsah M-H, Teuschl Y, Thrift AG, Tonelli M, Tovani-Palone MR, Tran BX, Tripathi M, Tsegaye GW, Ullah A, Unim B, Unnikrishnan B, Vakilian A, Valadan Tahbaz S, Vasankari TJ, Venketasubramanian N, Vervoort D, Vo B, Volovici V, Vosoughi K, Vu GT, Vu LG, Wafa HA, Waheed Y, Wang Y, Wijeratne T, Winkler AS, Wolfe CDA, Woodward M, Wu JH, Wulf Hanson S, Xu X, Yadav L, Yadollahpour A, Yahyazadeh Jabbari SH, Yamagishi K, Yatsuya H, Yonemoto N, Yu C, Yunusa I, Zaman MS, Zaman SB, Zamanian M, Zand R, Zandifar A, Zastrozhin MS, Zastrozhina A, Zhang Y, Zhang Z-J, Zhong C, Zuniga YMH, Murray CJL (2021) Global, regional, and national burden of stroke and its risk factors, 1990–2019: a systematic analysis for the global burden of disease study 2019. Lancet Neurol 20(10):795–820. https://doi.org/10.1016/s1474-4422(21)00252-0

2. Feigin VL, Nguyen G, Cercy K, Johnson CO, Alam T, Parmar PG, Abajobir AA, Abate KH, Abd-Allah F, Abejie AN, Abyu GY, Ademi Z, Agarwal G, Ahmed MB, Akinyemi RO, Al-Raddadi R, Aminde LN, Amlie-Lefond C, Ansari H, Asayesh H, Asgedom SW, Atey TM, Ayele HT, Banach M, Banerjee A, Barac A, Barker-Collo SL, Bärnighausen T, Barregard L, Basu S, Bedi N, Behzadifar M, Béjot Y, Bennett DA, Bensenor IM, Berhe DF, Boneya DJ, Brainin M, Campos-Nonato IR, Caso V, Castañeda-Orjuela CA, Rivas JC, Catalá-López F, Christensen H, Criqui MH, Damasceno A, Dandona L, Dandona R, Davletov K, de Courten B, deVeber G, Dokova K, Edessa D, Endres M, Faraon EJA, Farvid MS, Fischer F, Foreman K, Forouzanfar MH, Gall SL, Gebrehiwot TT, Geleijnse JM, Gillum RF, Giroud M, Goulart AC, Gupta R, Gupta R, Hachinski V, Hamadeh RR, Hankey GJ, Hareri HA, Havmoeller R, Hay SI, Hegazy MI, Hibstu DT, James SL, Jeemon P, John D, Jonas JB, Jóźwiak J, Kalani R, Kandel A, Kasaeian A, Kengne AP, Khader YS, Khan AR, Khang YH, Khubchandani J, Kim D, Kim YJ, Kivimaki M, Kokubo Y, Kolte D, Kopec JA, Kosen S, Kravchenko M, Krishnamurthi R, Kumar GA, Lafranconi A, Lavados PM, Legesse Y, Li Y, Liang X, Lo WD, Lorkowski S, Lotufo PA, Loy CT, Mackay MT, Abd El Razek HM, Mahdavi M, Majeed A, Malekzadeh R, Malta DC, Mamun AA, Mantovani LG, Martins SCO, Mate KK, Mazidi M, Mehata S, Meier T, Melaku YA, Mendoza W, Mensah GA, Meretoja A, Mezgebe HB, Miazgowski T, Miller TR, Ibrahim NM, Mohammed S, Mokdad AH, Moosazadeh M, Moran AE, Musa KI, Negoi RI, Nguyen M, Nguyen QL, Nguyen TH, Tran TT, Nguyen TT, Anggraini Ningrum DN, Norrving B, Noubiap JJ, O'Donnell MJ, Olagunju AT, Onuma OK, Owolabi MO, Parsaeian M, Patton GC, Piradov M, Pletcher MA, Pourmalek F, Prakash V, Qorbani M, Rahman M, Rahman MA, Rai RK, Ranta A, Rawaf D, Rawaf S, Renzaho AM, Robinson SR, Sahathevan R, Sahebkar A, Salomon JA, Santalucia P, Santos IS, Sartorius B, Schutte AE, Sepanlou SG, Shafieesabet A, Shaikh MA, Shamsizadeh M, Sheth KN, Sisay M, Shin MJ, Shiue I, Silva DAS, Sobngwi E, Soljak M, Sorensen RJD, Sposato LA, Stranges S, Suliankatchi RA, Tabarés-Seisdedos R, Tanne D, Nguyen CT, Thakur JS, Thrift AG, Tirschwell DL, Topor-Madry R, Tran BX, Nguyen LT, Truelsen T, Tsilimparis N, Tyrovolas S, Ukwaja KN, Uthman OA, Varakin Y, Vasankari T, Venketasubramanian N, Vlassov VV, Wang W, Werdecker A, Wolfe CDA, Xu G, Yano Y, Yonemoto N, Yu C, Zaidi Z, El Sayed Zaki M, Zhou M, Ziaeian B, Zipkin B, Vos T, Naghavi M, Murray CJL, Roth GA (2018) Global, regional, and country-specific lifetime risks of stroke, 1990 and 2016. N Engl J Med 379(25):2429–2437. https://doi.org/10.1056/NEJMoa1804492

3. Iadecola C, Buckwalter MS, Anrather J (2020) Immune responses to stroke: mechanisms, modulation, and therapeutic potential. J Clin Invest 130(6):2777–2788. https://doi.org/10.1172/jci135530

4. Feng L, Han C-X, Cao S-Y, Zhang H-M, Wu G-Y (2020) Deficits in motor and cognitive functions in an adult mouse model of hypoxia-ischemia induced stroke. Sci Rep 10(1):20646. https://doi.org/10.1038/s41598-020-77678-8

5. Kim Y, Khose S, Zaidat OO, Hassan AE, Fifi JT, Nanda A, Atchie B, Woodward B, Doerfler A, Tomasello A, Yoo AJ, Sheth SA (2022) Duration of ischemia affects outcomes independent of infarct size in stroke. Stroke Vasc Interv Neurol 2(5):e000163. https://doi.org/10.1161/SVIN.121.000163

6. Zhou Y, Zhang L, Ospel J, Goyal M, McDonough R, Xing P, Li Z, Zhang X, Zhang Y, Zhang Y, Hong B, Xu Y, Huang Q, Li Q, Yu Y, Zuo Q, Ye X, Yang P, Liu J, DIRECT-MT Investigators (2022) Association of intravenous alteplase, early reperfusion, and clinical outcome in patients with large vessel occlusion stroke: post hoc analysis of the randomized DIRECT-MT Trial. Stroke

53(6):1828–1836. https://doi.org/10.1161/STROKEAHA.121.037061

7. Tanaka Y, Nakagomi N, Doe N, Nakano-Doi A, Sawano T, Takagi T, Matsuyama T, Yoshimura S, Nakagomi T (2020) Early reperfusion following ischemic stroke provides beneficial effects, even after lethal ischemia with mature neural cell death. Cells 9 (6). doi: https://doi.org/10.3390/cells9061374
8. Alper BS, Foster G, Thabane L, Rae-Grant A, Malone-Moses M, Manheimer E (2020) Thrombolysis with alteplase 3-4.5 hours after acute ischaemic stroke: trial reanalysis adjusted for baseline imbalances. BMJ. Evid Based Med 25(5):168–171. https://doi.org/10.1136/bmjebm-2020-111386
9. Saver JL, Goyal M, van der Lugt A, Menon BK, Majoie CBLM, Dippel DW, Campbell BC, Nogueira RG, Demchuk AM, Tomasello A, Cardona P, Devlin TG, Frei DF, du Mesnil de Rochemont R, Berkhemer OA, Jovin TG, Siddiqui AH, van Zwam WH, Davis SM, Castaño C, Sapkota BL, Fransen PS, Molina C, van Oostenbrugge RJ, Chamorro Á, Lingsma H, Silver FL, Donnan GA, Shuaib A, Brown S, Stouch B, Mitchell PJ, Davalos A, Roos YBWEM, Hill MD, HERMES Collaborators (2016) Time to treatment with endovascular thrombectomy and outcomes from ischemic stroke: a meta-analysis. JAMA 316(12):1279–1289. https://doi.org/10.1001/jama.2016.13647
10. Fugate JE, Rabinstein AA (2015) Absolute and relative contraindications to IV rt-PA for acute ischemic stroke. Neurohospitalist 5(3):110–121. https://doi.org/10.1177/1941874415578532
11. Goyal M, Almekhlafi M, Dippel DW, Campbell BCV, Muir K, Demchuk AM, Bracard S, Davalos A, Guillemin F, Jovin TG, Menon BK, Mitchell PJ, Brown S, White P, Majoie CBLM, Saver JL, Hill MD, null n (2019) Rapid alteplase administration improves functional outcomes in patients with stroke due to large vessel occlusions. Stroke 50 (3):645–651. doi:https://doi.org/10.1161/STROKEAHA.118.021840
12. Evans MA, Lim R, Kim HA, Chu HX, Gardiner-Mann CV, Taylor KWE, Chan CT, Brait VH, Lee S, Dinh QN, Vinh A, Phan TG, Srikanth VK, Ma H, Arumugam TV, Fann DY, Poh L, Hunt CPJ, Pouton CW, Haynes JM, Selemidis S, Kwan W, Teo L, Bourne JA, Neumann S, Young S, Gowing EK, Drummond GR, Clarkson AN, Wallace EM, Sobey CG, Broughton BRS (2018) Acute or delayed systemic administration of human amnion epithelial cells improves outcomes in experimental stroke. Stroke 49(3):700–709. https://doi.org/10.1161/STROKEAHA.117.019136
13. Milani D, Clark VW, Cross JL, Anderton RS, Knuckey NW, Meloni BP (2016) Poly-arginine peptides reduce infarct volume in a permanent middle cerebral artery rat stroke model. BMC Neurosci 17(1):19. https://doi.org/10.1186/s12868-016-0253-z
14. Hetze S, Engel O, Römer C, Mueller S, Dirnagl U, Meisel C, Meisel A (2013) Superiority of preventive antibiotic treatment compared with standard treatment of poststroke pneumonia in experimental stroke: a bed to bench approach. J Cereb Blood Flow Metab 33(6):846–854. https://doi.org/10.1038/jcbfm.2013.6
15. Clarkson AN, Huang BS, Macisaac SE, Mody I, Carmichael ST (2010) Reducing excessive GABA-mediated tonic inhibition promotes functional recovery after stroke. Nature 468(7321):305–309. https://doi.org/10.1038/nature09511
16. Watson BD, Dietrich WD, Busto R, Wachtel MS, Ginsberg MD (1985) Induction of reproducible brain infarction by photochemically initiated thrombosis. Ann Neurol 17(5):497–504. https://doi.org/10.1002/ana.410170513
17. Knezic A, Broughton BRS, Widdop RE, McCarthy CA (2022) Optimising the photothrombotic model of stroke in the C57BI/6 and FVB/N strains of mouse. Sci Rep 12(1):7598. https://doi.org/10.1038/s41598-022-11793-6
18. Porritt MJ, Andersson HC, Hou L, Nilsson Å, Pekna M, Pekny M, Nilsson M (2012) Photothrombosis-induced infarction of the mouse cerebral cortex is not affected by the Nrf2-activator sulforaphane. PLoS One 7(7):e41090. https://doi.org/10.1371/journal.pone.0041090
19. Kretz CA, Vaezzadeh N, Gross PL (2010) Tissue factor and thrombosis models. Arterioscler Thromb Vasc Biol 30(5):900–908. https://doi.org/10.1161/ATVBAHA.108.177477
20. Bergeron M (2003) Inducing photochemical cortical lesions in rat brain. Curr Protoc Neurosci chapter 9:unit 9.16. https://doi.org/10.1002/0471142301.ns0916s23
21. Wester P, Watson BD, Prado R, Dietrich WD (1995) A photothrombotic 'ring' model of rat stroke-in-evolution displaying putative penumbral inversion. Stroke 26(3):444–450. https://doi.org/10.1161/01.STR.26.3.444
22. Clarkson AN, López-Valdés HE, Overman JJ, Charles AC, Brennan KC, Thomas Carmichael

S (2013) Multimodal examination of structural and functional remapping in the mouse photothrombotic stroke model. J Cereb Blood Flow Metab 33(5):716–723. https://doi.org/10.1038/jcbfm.2013.7

23. Ong LK, Zhao Z, Kluge M, Walker FR, Nilsson M (2017) Chronic stress exposure following photothrombotic stroke is associated with increased levels of amyloid beta accumulation and altered oligomerisation at sites of thalamic secondary neurodegeneration in mice. J Cereb Blood Flow Metab 37(4):1338–1348. https://doi.org/10.1177/0271678x16654920

24. Choi YK, Urnukhsaikhan E, Yoon HH, Seo YK, Park JK (2016) Effect of human mesenchymal stem cell transplantation on cerebral ischemic volume-controlled photothrombotic mouse model. Biotechnol J 11(11): 1397–1404. https://doi.org/10.1002/biot.201600057

25. Houlton J, Zubkova OV, Clarkson AN (2022) Recovery of post-stroke spatial memory and Thalamocortical connectivity following novel Glycomimetic and rhBDNF treatment. Int J Mol Sci 23(9). https://doi.org/10.3390/ijms23094817

26. Rennert RC, Wali AR, Steinberg JA, Santiago-Dieppa DR, Olson SE, Pannell JS, Khalessi AA (2019) Epidemiology, natural history, and clinical presentation of large vessel ischemic stroke. Neurosurgery 85(suppl_1):S4–S8

27. Baird AE, Donnan GA, Austin MC, Fitt GJ, Davis SM, McKay WJ (1994) Reperfusion after thrombolytic therapy in ischemic stroke measured by single-photon emission computed tomography. Stroke 25(1):79–85. https://doi.org/10.1161/01.str.25.1.79

28. Clark TA, Sullender C, Kazmi SM, Speetles BL, Williamson MR, Palmberg DM, Dunn AK, Jones TA (2019) Artery targeted photothrombosis widens the vascular penumbra, instigates peri-infarct neovascularization and models forelimb impairments. Sci Rep 9(1):2323. https://doi.org/10.1038/s41598-019-39092-7

29. Kawano H, Bivard A, Lin L, Ma H, Cheng X, Aviv R, O'Brien B, Butcher K, Lou M, Zhang J, Jannes J, Dong Q, Levi CR, Parsons MW (2016) Perfusion computed tomography in patients with stroke thrombolysis. Brain 140(3):684–691. https://doi.org/10.1093/brain/aww338

30. Liu Y, Li Y, Zang J, Zhang T, Li Y, Tan Z, Ma D, Zhang T, Wang S, Zhang Y, Huang L, Wu Y, Su X, Weng Z, Deng D, Tsang CK, Xu A, Lu D (2022) CircOGDH is a penumbra biomarker and therapeutic target in acute ischemic stroke. Circ Res 130(6):907–924. https://doi.org/10.1161/CIRCRESAHA.121.319412

31. Peña-Martínez C, Durán-Laforet V, García-Culebras A, Ostos F, Hernández-Jiménez M, Bravo-Ferrer I, Pérez-Ruiz A, Ballenilla F, Díaz-Guzmán J, Pradillo JM, Lizasoain I, Moro MA (2019) Pharmacological modulation of neutrophil extracellular traps reverses thrombotic stroke tPA (tissue-type plasminogen activator) resistance. Stroke 50(11): 3228–3237. https://doi.org/10.1161/strokeaha.119.026848

32. Alonso-Alonso ML, Pérez-Mato M, Sampedro-Viana A, Correa-Paz C, Ávila-Gómez P, Sobrino T, Campos F, Castillo J, Iglesias-Rey R, Hervella P (2022) Fibrin-targeted nanoparticles for finding, visualizing and characterizing blood clots in acute ischemic stroke. Pharmaceutics 14(10). https://doi.org/10.3390/pharmaceutics14102156

33. Sun YY, Kuo YM, Chen HR, Short-Miller JC, Smucker MR, Kuan CY (2020) A murine photothrombotic stroke model with an increased fibrin content and improved responses to tPA-lytic treatment. Blood Adv 4(7):1222–1231. https://doi.org/10.1182/bloodadvances.2019000782

34. Kuo YM, Sun YY, Kuan CY (2021) A fibrin-enriched and tPA-sensitive photothrombotic stroke model. J Vis Exp 172. https://doi.org/10.3791/61740

35. Iannucci J, Renehan W, Grammas P (2020) Thrombin, a mediator of coagulation, inflammation, and neurotoxicity at the neurovascular Interface: implications for Alzheimer's disease. Front Neurosci 14:762. https://doi.org/10.3389/fnins.2020.00762

36. Franklin KB (1997) PG the mouse brain in stereotaxic coordinates. Academic Press, San Diego

Corrections to: Neurobiology: Methods and Protocols

Sebastian Dworkin

Corrections to:
Chapters 3, 6 & 15 in: Sebastian Dworkin (ed.), *Neurobiology: Methods and Protocols*, Methods in Molecular Biology, vol. 2746,
https://doi.org/10.1007/978-1-0716-3585-8

In the original version of this book,

- In Chapter 3 "Mechanical Ablation of Larval Zebra Fish Spinal Cord", the term "zebrafish" was misspelt as "zebra fish" which has been rectified along with a few other contextual amendments.
- In Chapter 6 "In Situ Hybridization to Characterize Neurulation and Midbrain–Hindbrain Boundary Formation in Zebra Fish", the term "zebrafish" was misspelt as "zebra fish" which has been rectified
- In Chapter 15 "Active Induction of a Multiple Sclerosis-Like Disease in Common Laboratory Mouse Strains", Figures 8 and 9 were placed in incorrect order, which has been rectified.

The updated version of these chapters can be found at
https://doi.org/10.1007/978-1-0716-3585-8_3
https://doi.org/10.1007/978-1-0716-3585-8_6
https://doi.org/10.1007/978-1-0716-3585-8_15

Sebastian Dworkin (ed.), *Neurobiology: Methods and Protocols*, Methods in Molecular Biology, vol. 2746,
https://doi.org/10.1007/978-1-0716-3585-8_19,

INDEX

Sebastian Dworkin (ed.), *Neurobiology: Methods and Protocols*, Methods in Molecular Biology, vol. 2746, https://doi.org/10.1007/978-1-0716-3585-8,

MIX
Papier aus verantwortungsvollen Quellen
Paper from responsible sources
FSC® C105338
FSC
www.fsc.org

If you have any concerns about our products,
you can contact us on
ProductSafety@springernature.com

In case Publisher is established outside the EU,
the EU authorized representative is:
Springer Nature Customer Service Center GmbH
Europaplatz 3, 69115 Heidelberg, Germany

Printed by Libri Plureos GmbH
in Hamburg, Germany